Pervasive Computing

Concepts, Technologies and Applications

T0094230

Pervasive Computing

Concepts, Technologies and Applications

Minyi Guo • Jingyu Zhou
Feilong Tang • Yao Shen

CRC Press
Taylor & Francis Group
Boca Raton London New York

CRC Press is an imprint of the
Taylor & Francis Group, an **informa** business

CRC Press
Taylor & Francis Group
6000 Broken Sound Parkway NW, Suite 300
Boca Raton, FL 33487-2742

© 2017 by Taylor & Francis Group, LLC
CRC Press is an imprint of Taylor & Francis Group, an Informa business

No claim to original U.S. Government works

Printed on acid-free paper
Version Date: 20160609

International Standard Book Number-13: 978-1-4665-9627-6 (Hardback)

Library of Congress Cataloging-in-Publication Data

Names: Guo, Minyi. | Zhou, Jingyu. | Tang, Feilong. | Shen, Yao, (Computer scientist)
Title: Pervasive computing : concepts, technologies and applications / authors, Minyi Guo, Jingyu Zhou, Feilong Tang, and Yao Shen.
Description: Boca Raton : Taylor & Francis, CRC Press, 2017. | Includes bibliographical references.
Identifiers: LCCN 2016010722 | ISBN 9781466596276
Subjects: LCSH: Ubiquitous computing.
Classification: LCC QA76.5915 .G86 2107 | DDC 004--dc23
LC record available at https://lccn.loc.gov/2016010722

**Visit the Taylor & Francis Web site at
http://www.taylorandfrancis.com**

**and the CRC Press Web site at
http://www.crcpress.com**

Printed and bound in the United States of America by Publishers Graphics, LLC on sustainably sourced paper.

Contents

Preface

Our research group in the Embedded and Pervasive Computing Laboratory at Shanghai Jiao Tong University began working on various pervasive computing projects in 2006. At that time, smartphones had not been invented, and cloud computing had yet to gain momentum. To realize the focal point of pervasive computing (i.e., computing anywhere and anytime), we mainly used a combination of personal computers, personal digital assistants (PDAs), and small sensors. Because we work and study primarily at the university, the pervasive applications developed over the years focus on a campus environment for providing relevant services, such as a campus map, short message service (SMS), photo sharing, and a campus search.

Looking back, many of the services and tools we developed are now becoming key parts of features provided by modern smartphones. However, the underlying technology may still be valuable for future endeavors. Over the last few years, industry and academia have developed the techniques of cloud computing and mobile computing to realize pervasive computing perspectives. Today, a smartphone can help us do almost everything in our daily lives: booking tickets, ordering takeout, making a purchase, watching TV, playing a game, and so on. These implementations are based not only on the rapid increases in performance and speed of hardware and communication infrastructures but also on the key software and middleware techniques for pervasive and mobile computing.

This book is organized into eight chapters. The first two chapters introduce pervasive computing concepts and the structure of pervasive computing systems. In Chapter 1, we first introduce the concepts and perspectives of pervasive computing, then we list some challenges for implementing a pervasive computing system. Several key technologies are then introduced such as middleware, context awareness, resource management, human–computer interaction (HCI), pervasive transaction processing, and user preference and recommendation. In Chapter 2 through Chapter 7, different technological aspects of pervasive computing are discussed. Chapter 3 describes the aspects of context awareness. Chapter 4 gives the methods showing how to allocate resources efficiently and how to migrate tasks intelligently. An HCI migration framework is proposed in Chapter 5. Many pervasive applications could be used in pervasive transaction processing, such as mobile bank transfers. In Chapter 6, we present a context-aware transaction model and propose a context-adaptive dynamical transaction management algorithm. Recommendation is a way to help people find information for their needs and is widely used in many online services for suggesting to customers the products they might like to buy. Chapter 7 focuses on user preferences and recommendation systems. Finally, Chapter 8 outlines two case studies.

This book is written in the hope that, by presenting various aspects of pervasive computing, readers are introduced not only to key concepts but also to various techniques and typical applications. Thus, a new generation of pervasive computing hackers will be freed from the need to slowly reinvent old wheels and will instead be able to focus on new and challenging frontiers during their journey. It is hoped that this book will be useful for the reader and that this exposure to pervasive computing might bring the reader as much fun and excitement—and as many challenges—as it has given the authors over the years.

Finally, we thank many past members of our group who participated in pervasive computing projects. Without them, this book would never have been the same. They are Mianxiong Dong, Shiwei Hu, Hu Guan, Huakang Li, Long Zheng, Yifei Wang, Daqiang Zhang, Hao Zhou, Min Wang, Minjie Wang, Xiaoxin Tang, Cansheng Ji, Yunlong Zhang, Jia Cheng, and Linchun Cao.

Minyi Guo, Jingyu Zhou, Feilong Tang, and Yao Shen
Shanghai Jiao Tong University
Shanghai, China

About the Authors

 **Minyi Guo** received his PhD degree in information science from University of Tsukuba, Japan in 1998. Currently he is the Zhiyuan Chair professor and chair of the Department of Computer Science and Engineering, Shanghai Jiao Tong University (SJTU), Shanghai, China. Before joining SJTU, Dr. Guo was a professor of the School of Computer Science and Engineering, University of Aizu, Aizu-Wakamatsu, Japan. Dr. Guo was awarded the National Science Fund award for Distinguished Young Scholars from the National Natural Science Foundation of Science (NSFC) in 2007 and was supported by the Recruitment Program of Global Experts in 2010. His present research interests include parallel/distributed computing, compiler optimizations, embedded systems, pervasive computing, and cloud computing. He has more than 300 articles in major journals and international conferences in these areas. Dr. Guo received five "best paper" awards from international conferences. He was associate editor of *IEEE Transactions on Parallel and Distributed Systems* and *IEEE Transactions on Computers*. Dr. Guo is a senior member of Institute of Electrical and Electronics Engineers (IEEE) and is a member of Association for Computing Machinery (ACM) and China Computer Federation (CCF).

 Jingyu Zhou received his PhD degree in computer science from University of California, Santa Barbara in 2006. He is currently an associate professor at Shanghai Jiao Tong University, Shanghai, China. He is interested in distributed systems, information retrieval, and security. He has published more than 40 papers at various conferences and in journals, including International Conference on World Wide Web, ACM/IEEE Conference on High Performance Networking and Computing, IEEE Transactions on Parallel and Distributed Systems, IEEE Transactions on Big Data, and International Conference on Dependable Systems and Networks. He has served as a Program Committee member for more than 20 international conferences.

Feilong Tang received his doctoral degree in computer science from Shanghai Jiao Tong University (SJTU), Shanghai, China, in 2005. Currently, he is a full professor in the Department of Computer Science and Engineering at SJTU. He previously was the JSPS (Japan Society for the Promotion of Science) Research Fellow, and he received the Distinguished Pu-Jiang Scholars Award from Shanghai Municipality. His research interests focus on mobile cognitive networks, clouding computing, and big data. He has received two best papers from international conferences. Dr. Tang is the Institution of Engineering and Technology (IET) Fellow and is a member of Institute of Electrical and Electronics Engineers (IEEE) and Association for Computing Machinery (ACM). He has served as program cochair for eight international conferences (workshops).

Yao Shen received his PhD degree in computer science and engineering from Shanghai Jiao Tong University (SJTU), Shanghai, China, in 2007. Currently, he is an associate professor in computer science at Shanghai Jiao Tong University (SJTU), Shanghai, China. He has published extensively in the areas of distributed computing, cloud computing, and computer networking, and obtained more than ten patents. He received the 2010 and the 2013 Shanghai Science and Technology Award. Dr. Yao Shen pays much attention to system implementation while doing research. He won the Intel Prize in the 2011 Open Source Software World Challenge.

Chapter 1

Pervasive Computing Concepts

Since Mark Weiser first coined the phrase *ubiquitous computing* in 1988, when he was the chief technology officer at Xerox's Palo Alto Research Center (PARC), the development of pervasive/ubiquitous computing has continued for more than quarter century. During this period, the popularization of mobile Internet devices such as smartphones and smart pads has led to the core perspective of pervasive computing (that of computing devices being available everywhere at any time) becoming the main trend in computing and information technology. As part of our daily lives, these devices help us connect to worldwide networks without boundaries and provide us with quick and secure access to a wealth of information and services.

This book aims to introduce a foundation of the concepts, architecture, key techniques, and typical applications of pervasive computing. The main body of this book focuses on context awareness, resource management, human–computer interface, pervasive transaction processing, and user preference discovery—all of which have been previously researched by our team.

1.1 Perspectives of Pervasive Computing

To help convey the *look and feel* of such a world, we present two hypothetical scenarios.

The first scenario involves a traveler. Dr. John is driving his car on a long distance trip. Along the way, the car detects when fuel will be used up, and it then automatically searches for gas stations nearby. In a short time, the car finds an appropriate location that can provide fuel at the lowest price among gas stations within 1 km. The pervasive device displays a suggestion to John and guides him to the selected gas station. Two hours later, it is lunch time and John is hungry. The pervasive devices search for an appropriate restaurant near John's current location, according to his food preferences.

Our second scenario focuses on daily life. Professor Li is watching a video in his office using a personal computer (PC) when he receives a call asking him to attend a meeting. On his way there, John continues to watch the video using his smartphone. A pervasive computing system will recover the point in the video where John stopped watching it on his PC and will automatically download a low-resolution video to John's smartphone (which has a low-resolution screen), connecting with the video server through a low-bandwidth wireless channel. In this case, the complex switch process is completely unnoticed by Professor Li.

From these scenarios, we observe that the whole is much greater than the sum of its parts. In other words, the real research takes place with the seamless integration of component technologies into a system such as iShadow [1].

1.1.1 Technology Trend Overview

The advancement of mobile computing, cloud computing, wearable computing, and smart devices has brought unprecedented opportunities for pervasive computing. In recent years, we have seen a sharp rise in the shipment of smartphones and smart pads. These mobile devices are equipped with multicore central processing units (CPUs), gigabytes of memory, Wi-Fi and cellular communication components, a global positioning system (GPS), and other sensors. As a result, in the not too distant future, they will have more powerful information storage and processing capabilities than desktop computers. In the meantime, millions of apps have been developed for these mobile devices that provide a wide variety of personalized services.

Previous pervasive computing was often limited by the physical size of the electronic devices. However, this constraint has been overcome by rapid technological developments. A variety of wearable devices have entered the mass market, such as Google Glass, Jawbone UP [2], Nike FuelBand [3], and Sony SmartWatch [4]. These wearable devices have successfully extended computation, storage, and communication capabilities to personal gadgets, providing a unique opportunity for expanding the sensing capabilities of users and offering situation-aware information. For instance, a user with a Google Glass may visualize the transit route of nearby metro stations within the device's screen.

Additionally, many traditional appliances are now embedded with sensors and communication components, thus making them *smarter*. For instance, the Smart Sofa [5] can identify the people sitting on it using programmable sensors in the sofa legs and can address users with personalized greetings.

1.1.2 Pervasive Computing: Concepts

Pervasive computing is a new wave of technology where computing takes place everywhere and anywhere. The goal of pervasive computing is to become a *technology that disappears* [6]. Pervasive computing can be seen as an enabler technology for many new and exciting applications, making information accessible to anyone, anywhere, and at any time [7].

Pervasive computing saw rapid growth in the past few decades due to a shift from a technology perspective to a utility and usability perspective. Pervasive computing, although currently used in today's fast-paced lifestyle, will also be the future of this transition [8].

Pervasive computing projects have emerged at major universities and in industry. For example, Carnegie Mellon University, the University of California at Berkeley (UC Berkeley), Massachusetts Institute of Technology (MIT), and the University of Washington set up the Aura, Endeavour, Oxygen, and Portolano projects, respectively. These represent a broad communal effort to make pervasive computing a reality, focusing on technology that is more social and more people oriented [6].

1.2 Challenges

Although pervasive computing offers a lot of new, interesting, and useful possibilities, many challenges need to be overcome before the vision becomes real. The main challenges faced in pervasive computing are as follows [6]:

1. *Transparency.* Pervasive computing is people-oriented, providing transparent services based on individual requirements, preferences, and so on. In contrast to desktop computing, pervasive computing can use any device, in any location, and in any format, without an individual being consciously aware of what is taking place with respect to the actual computing.
2. *Context awareness.* Typically, pervasive computing systems are very tightly connected with specific users. Pervasive computing systems often gather and store information on the user's behavior, context, habits, and planning. This information forms the basis for the many benefits the system can offer individual users.
3. *Mobility.* The omnipresence of ubiquitous applications typically is achieved by either having devices move with the user or by having applications move among mobile users' devices. In both cases, applications need to adapt to the moving environment, which involves maintaining device connections and adapting protocol for handling mobility. Although some of the problems can be addressed by routing and handovers, many cannot be solved at the network level because knowledge of application semantics is required for runtime adaptation.

4. *Heterogeneous devices.* Pervasive applications typically involve many different types of devices working in an orchestrated way. As a user moves around in an environment, the servicing application often moves with him. For instance, the application may switch from a desktop PC to a mobile phone. In addition, the heterogeneous devices complicate the development of high-level applications because different devices provide varied programming interfaces, resource abstractions, and functional capabilities.

5. *Data management.* In a pervasive environment, many devices continuously produce huge amounts of raw data. It is challenging to transfer, store, and process these data. In order to provide meaningful semantics for high-level applications, some preprocessing of the data may be necessary. For instance, inaccurate data can cause context management systems to produce false context information, which can result in incorrect reasoning for applications.

6. *Fault tolerance.* The pervasive application requires harmonious cooperation among many hardware devices and various software components. However, hardware devices such as sensors are prone to failures, and software components can often malfunction. Both types of failures can hamper the effectiveness of pervasive applications. Thus, fault tolerance measures must be built into the system to avoid serious or fatal consequences for users.

7. *Reliability.* Pervasive computing systems normally are embedded in the environment not only for gathering information but also for making decisions or at least for decision support. Limited resources and unstable wireless communication can impact the reliability of pervasive computing, making its application a more arduous task.

8. *Usability.* Pervasive computing means that smart and agile computing devices, though invisible, are embedded everywhere in the environment. Therefore, usability is the critical success factor for these applications, making this the hottest research topic in the field of pervasive computing. An easy-to-use and intuitive interface is the demand of the future [9]. Ease of use will follow different levels for different end users. Usability in pervasive computing applications is very important and requires more attention than all other areas, such as desktop and Web-based applications.

1.3 Technology

As noted in Section 1.1.1, pervasive computing involves state-of-the-art technology—such as a middleware and programming model, context-awareness computing, resource allocation and management, human–computer interaction (HCI), pervasive transaction processing, and data mining. We briefly discuss these aspects here.

1.3.1 Middleware

Pervasive applications often rely on some middleware support. This is not a coincidence. Application programmers face a large variety of challenges, such as data management, context management, security and privacy issues, and mobility. Solving these problems for each application is error prone and inefficient. Thus, it is more desirable to extract the common requirements from different applications and provide the desired functionality from a middleware layer, which hides the complexity and heterogeneity of underlying hardware and network platforms.

The pervasive middleware primarily addresses issues from high-level applications on one hand while dealing with the complexity of operating underlying devices, networks, and platforms on the other. High-level application requirements are often diverse and application specific, with the common themes being context awareness, locality, reliability, adaptability, and reusability.

The operations of pervasive middleware are realized by heterogeneous underlying devices, networks, and platforms. In practice, these operations are also constrained by physical resources, such as low computation capability and insufficient battery supply. As a result, the middleware must strike a balance between functionality provided for the application layer and the resource usage of underlying hardware devices. In different scenarios, one aspect of consideration can outweigh the other and can affect design decisions.

1.3.2 Context Awareness

Context refers to the pieces of information that capture the characteristics of pervasive computing environments [9,10]. We classify these into physical and virtual contexts based on context sources [11]. Physical contexts may be aggregated by physical sources, for example, sensing and computational devices such as handheld devices, wireless sensors, and Radio Frequency Identification (RFID), which involve accelerated speed, air pressure, light, location, movement, sound, touch, and temperature. Virtual contexts are specified by users or captured from user interactions—including user preferences, business processes, goals, and tasks.

Context awareness is a mechanism that assists pervasive applications in adapting their behaviors to the evolving contexts [12]. Suppose a call comes when Alice is watching *Kung Fu Panda* in her smart bedroom. Depending upon the urgency of the call and the caller's relationship with Alice, this smart space adapts its behavior correspondingly (i.e., it will reject or accept the call). Most pervasive applications achieve context awareness in a similar fashion. They first acquire physical and virtual contexts and then exploit these contexts to determine what strategy should be taken when contexts keep evolving.

This book will cover the physical and virtual aspects through discussions of wireless sensor networking, user tracking, and context reasoning. Using wireless sensors is a basic, common, and effective way to gather physical contexts for pervasive applications. In a smart space, the architectural design of a wireless sensor network is a key challenge for achieving scalability, robustness, and balanced energy dissipation of the network. Location information is the most important context in pervasive computing. Accurate user tracking facilitates the provision of service in smart spaces. Based on context information, reasoning enables context awareness for pervasive applications. In this book, evidence theory, the Dempster–Shafer theory of context reasoning, evidence propagation, and evidence selection strategy will be discussed extensively.

1.3.3 Resource Management

Pervasive devices exhibit very different features and functions. Resources in pervasive systems are limited and heterogeneous. Therefore, efficient resource management policies play a very important role in pervasive computing. In addition, task migration is another important issue related to resource allocation. More specifically, whenever tasks need to be migrated from one networked node to another, some resources (e.g., memory) need to be reallocated and others (e.g., some data) will be moved.

In this book, we will present several efficient resource allocation solutions—for example, pipeline-based resource allocation and probabilistic approach-based resource allocation algorithms for improving the performance of pervasive systems.

Moreover, we will present an intelligent task migration platform, xMozart, outlining the following technical aspects: application reconstruction, application state recovery, resource rebinding, and input/output (I/O) interface reinstallation for migrating applications in and between intelligent computing contexts. In xMozart, we migrate applications but do not migrate resources, and we try to minimize the connections between migrated and original applications. The goal is to migrate applications seamlessly—like a shadow—as users change their locations in certain contexts while demanding access to applications that originally resided in their main workstations.

1.3.4 Human–Computer Interaction

HCI provides channels for people to communicate with digital worlds. In smart spaces, HCI marks itself off from traditional interaction with various computing devices (e.g., TV, smartphone, refrigerator, etc.) by incorporating diversified interaction modalities (e.g., speech, gesture, face expression, etc.). Therefore, pervasive applications with the objective of providing human-centered services essentially require HCI services that are able to migrate their interaction across different devices or modalities in pervasive environments. We define this capability as interaction migration.

To achieve interaction migration in smart spaces, a Web service-based HCI migration framework will be introduced in which interaction application logic is modeled as interaction service and user preference, and context awareness is addressed by incorporation into descriptions of interaction interfaces. We discuss the HCI service selection process and algorithm that consider not only context information and user preferences but also interservice relations.

1.3.5 *Pervasive Transaction Processing*

Pervasive computing is a user-centric, scalable, parallel, and distributed computing paradigm, allowing users to access their preferred services even while moving around. Many key pervasive applications need software, hardware, and network reliability to be hidden from users.

Transaction management for pervasive environments has to provide mobile users with reliable and transparent services anytime and anywhere. Due to the limited resources of pervasive devices, high mobility of users, and the transparency requirements of pervasive computing, however, traditional mobile transaction processing models cannot support pervasive applications effectively.

In this book, we will present a context-aware transaction model and a context-adaptive and energy-efficient transaction management mechanism (CETM) for pervasive transaction processing. This transaction processing model is able to adapt to changing environments dynamically, providing transparency for pervasive applications. The CETM can dynamically adjust transaction execution behaviors in terms of current context information, and it is able to significantly reduce the failure probability of concurrent pervasive transactions. On the other hand, considering the features of pervasive environments, we have designed a mesh network-based architecture for online pervasive transactions. This transaction processing architecture provides a breakthrough in the limitations of the traditional client/proxy/server framework, allowing users to access pervasive services anytime and anywhere.

1.3.6 *User Preference and Recommendations*

To better serve human users, pervasive applications often adapt their behavior according to different users' preferences. For instance, when a user reads news articles with an RSS reader, it is desirable for the RSS reader to know that the user is more interested in financial news than politics, thus giving financial articles a more prominent position in the whole list of news reports.

The key to such a smart reader lies in two important underlying principles. First, the smart reader needs a personal profile that captures the user's preferences across different topics. Second, the reader should reorganize a list of articles according to the user's preferences, making effective recommendations.

Generally, the user preference can be explicitly built by asking the user to fill in personal information. However, such an approach is burdensome, and the users may not be willing to explicitly specify their needs. As a result, many systems adopt an implicit approach to find user preferences by learning from past interactions with each individual user.

Pervasive applications often make recommendations for users in finding relevant information. Previous recommender systems can be categorized into content-based and collaborative filtering approaches. Content-based approaches recommend items to a user that are similar to past favorites. In contrast, collaborative filtering (CF) approaches do not look at specific item features. Instead, CF approaches make predictions based on large-scale item–user matrices. The idea is to exploit similarities among different users. If a user prefers both item A and B, then for a similar user who likes A, the system recommends B.

Finally, recommendation in social networks has many important applications. The social network is an important way of spreading information via *word-of-mouth*—the selected users influence their friends on the social network, and those friends then influence their friends, until finally a large number of users choose a particular product that has been recommended. Such a spread of information is an important marketing strategy. Some other important applications in the social network include finding the most important blogs and searching for domain experts. All of these applications can be categorized as recommendation applications.

References

1. D. Zhang, H. Guan, J. Zhou, F. Tang, and M. Guo, iShadow: Yet another pervasive computing environment, in *Proceedings of International Symposium on Parallel and Distributed Processing with Applications (ISPA)*, December 10–12, 2008, pp. 261–268.
2. Jawbone, Up. [Online]. Available from: https://jawbone.com/up, accessed on May 13, 2014.
3. Nike, FuelBand, 2014. [Online]. Available from: http://www.nike.com/us/en_us/c/nikeplus-fuelband, accessed on May 13, 2014.
4. Sony, SmartWatch, 2014. [Online]. Available from: http://www.sonymobile.com/us/products/accessories/smartwatch/, accessed on May 13, 2014.
5. J. Legon, 'Smart sofa' aimed at couch potatoes, *CNN*, 2003. [Online]. Available from: http://www.cnn.com/2003/TECH/ptech/09/22/smart.sofa/, accessed on May 13, 2014.
6. M. Satyanarayanan, Pervasive computing: Vision and challenges, *IEEE Personal Communications*, vol. 8, no. 4, pp. 10–17, 2001.
7. D. Zhang and B. Adipat, Challenges, methodologies, and issues in the usability testing of mobile applications, *International Journal of Human-Computer Interaction*, vol. 18, no. 3, pp. 293–308, 2005.
8. A. W. Muzaffar, F. Azam, H. Anwar, and A. S. Khan, Usability aspects in pervasive computing: Needs and challenges, *International Journal of Computer Applications*, vol. 32, no. 10, pp. 18–24, 2011.

9. C. Xu, S. C. Cheung, W. K. Chan, and C. Ye, Heuristics-based strategies for resolving context inconsistencies in pervasive computing applications, in *Proceedings of the 28th International Conference on Distributed Computing Systems (ICDCS 2008)*, June 17–20, 2008, pp. 709–717.
10. Y. Huang, X. Ma, J. Cao, X. Tao, and J. Lu, Concurrent event detection for asynchronous consistency checking of pervasive context, in *IEEE International Conference on Pervasive Computing and Communications*, March 9–13, 2009, pp. 131–139.
11. D. Zhang, H. Huang, C.-F. Lai, X. Liang, Q. Zou, and M. Guo, Survey on context-awareness in ubiquitous media, *Multimedia Tools and Applications*, vol. 67, no. 1, pp. 179–211, 2013.
12. Z. Lei and N. D. Georganas, Context-based media adaptation in pervasive computing, in *Canadian Conference on Electrical and Computer Engineering*, May 13–16, 2001, pp. 913–918.

Chapter 2

The Structure and Elements of Pervasive Computing Systems

There are many perspectives on pervasive computing. The key components, from our point of view, could be a platform that integrates the basic elements of a context-aware pervasive computing systems—from sensors and modeling to reasoning techniques. In this chapter, we discuss the structure of a pervasive computing system that we have developed. As illustrated in Figure 2.1, we divide our discussion into three layers: the infrastructure layer, the middleware layer, and the application layer. The infrastructure layer consists of the hardware devices and infrastructure for pervasive applications. For high-level applications, the hardware resources are abstracted and managed by the middleware layer. High-level applications, such as smart campus and smart car space [1], are built on top of the middleware layer and support different user activities.

2.1 Infrastructure and Devices

Pervasive computing environments consist of a large number of devices embedded everywhere in our living environment. The number of pervasive applications is only limited by our imagination. For instance, a refrigerator can place an order to the supermarket when an item runs out. Here, the infrastructure provides network access for the refrigerator to contact the supermarket. Thus, the infrastructure for pervasive computing contains a variety of devices. As shown in Figure 2.1,

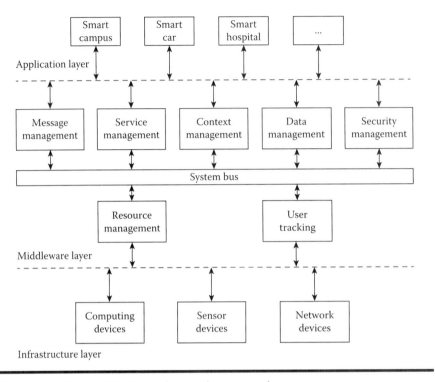

Figure 2.1   The architecture of pervasive computing systems.

we classify devices in the infrastructure layer into *computing devices, sensor devices,* and *network devices* according to their functionalities.

- *Computing devices* in a pervasive computing environment provide the capability of information processing. Traditional desktop machines, laptops, personal digital assistant (PDAs), and newer Pads and smartphones are examples of computing devices. Additionally, many objects (such as vehicles containing embedded CPUs) are also computing devices.
- *Sensors and actuators* represent devices that can sense or mediate between physical and virtual environments. Sensors collect information such as temperature, location, humidity, light, sound, and tension from the environment. As a result, different sensors are the enabling technologies for pervasive applications to achieve their specific goals and functionalities. Among various sensors, location sensors have commonly been built in many objects (e.g., mobile phones), and we have witnessed many new location-based pervasive applications are emerging in recent years, mainly due to the availability of location information from sensors.
- *Network devices* provide wired or wireless network access to connect different devices together in a virtual environment. Due to the variety of wireless

technologies, we see many devices using Bluetooth, Wi-Fi, infrared, Zigbee, long term evolution (LTE), near field communication (NFC), and radio frequency identification (RFID). The networking capability is essential in many scenarios where various pervasive devices can work in concert. For instance, a smart car assistant can use an LTE connection to enlist help from cloud-based voice recognition services to understand the voice command issued by the user.

As the size of sensors and communication chips becomes smaller, it is possible to integrate these devices into existing physical objects, such as embedding RFID tags into commodities in a supermarket or into books in a library. Thus, the borders of pervasive computing have been greatly expanded by the development of new technologies.

Finally, we note that many objects are very complex and may include partial or all of the aforementioned devices. For instance, a smartphone equipped with a CPU, memory, and storage is a computing device. Additionally, the phone contains a global positioning system (GPS) sensor and a Wi-Fi module. As a result, the phone is also a network device with sensors, and a focal point of pervasive research is to use these phones as application platforms. Another recent trend is to use wearable devices such as wristbands and watches [2–4].

2.1.1 Wireless Networks

The goal of pervasive computing is to create ambient intelligence. To make this vision a reality, interconnected wireless networks play a vital role in collecting context information, providing essential communication support for information flow, and in processing in a dynamically changing environment. Next, we discuss several types of emerging wireless networks.

In recent years, various interconnected wireless sensor networks (WSNs) have been set up to collect different environmental information and scattered objects. WSNs are ideal for many applications such as in the military, environment monitoring, intelligence control, traffic management, medical treatment, manufacture industry, and in antiterrorism activities. However, due to the limited power, computing, and memory of sensor nodes, WSNs suffer from unbalanced energy consumption among nodes, poor scalability, and single point of failure of sink nodes.

Due to the fundamental limitations of WSNs, wireless mesh sensor networks (WMSNs) are attracting more and more attention from both industry and academic communities to improve the scalability, reliability, and throughput of sensor networks and to support the node mobility. By deploying some super mesh nodes with powerful capacities to transmit data long distance, WMSNs take advantage of both mesh networks and WSNs. WMSNs not only provide the capacity to interconnect multiple heterogeneous sensor networks, but they also improve the scalability, robustness, and data throughput of sensor networks.

A wireless mesh network (WMN) is another wireless network architecture, which is self-organized, self-configured, and decentralized. When devices are added to or moved from networks, WMNs are able to automatically discover topology change and self-adaptively modify routing for more efficient data transmission. There are two kinds of nodes in WMNs: mesh routers and mobile clients. Equipped with powerful capacities and lower mobility, mesh routers automatically set up and maintain a wireless connection, forming the backbone of WMNs and possibly interconnecting with other kinds of networks (e.g., the Internet). Mobile clients can send and receive data. Traditional wireless devices have to first connect with an access point (AP) to communicate with other devices, even if they are located within radio range of each other. In contrast, each node in a WMN may directly communicate with neighboring nodes.

2.2 Middleware for Pervasive Computing Systems

The difficulty in developing pervasive applications lies in several aspects. First, pervasive applications use many devices of various types. Coordination and management of these devices are nontrivial tasks. Considering the resource constraints of these devices, applications may need to address frequent failures due to energy or memory size limitations. Second, sensor devices continuously collect a large amount of raw data about users and the environment. These raw data require efficient storage management and processing in order to provide meaningful context information for high-level applications. Sensor data are often noisy and conflicting, which requires data processing to filter most false events and keep all crucial events. High-level reasoning must be able to process events with tolerance for inconsistent low-level events. Third, the applications need context information in order to provide a better user experience. For instance, a personal health assistant needs to know if the user has had lunch so that it can choose to remind the user to take certain pills at a specific time. Finally, pervasive applications need to adapt to the moving environment, which involves the maintenance of connections with devices and protocol adaptation for handling mobility. Some of the problems can be solved by routing and handovers. However, many problems cannot be solved at the network level because knowledge of application semantics is required for runtime adaptation.

In order to address the aforementioned difficulties, the key enabling technology for pervasive computing applications is the middleware layer. As shown in Figure 2.1, the pervasive middleware is responsible for abstracting resources provided by underlying heterogeneous devices and providing a unified interface for high-level applications. The middleware hides the complexity and heterogeneity of underlying hardware and network platforms, eases the management of system resources, and makes the execution of applications more predictable.

There have been a number of pervasive middleware systems, such as Gaia [5], Aura [6], PICO/SeSCo [7], CORTEX [8], and Scenes [9]. These middlewares provide different abstractions and system support for applications. For instance, only Gaia provides user authentication and access control as security mechanism. Gaia, Aura, and CORTEX have context management, while PICO/SeSCo does not. In the following AWXROIN, we discuss the functionality of the middleware layer, including resource management, user tracking, context management, service management, data management, and security.

2.2.1 Resource Management

Pervasive applications require a seamless orchestration of a large body of heterogeneous sensors, devices, components, and services that may dynamically join or leave. For instance, the Computers in the Human Interaction Loop (CHIL) project [10] deals with a large number of perceptual sensors, such as 2D and 3D visual sensors, acoustic sensors, and audiovisual sensors. To better support these applications, pervasive middleware needs to efficiently manage sensors and actuating devices, despite the diverse functionality of underlying technologies and vendors. Key functions of resource management include *resource registration, resource control*, and *resource abstraction*.

Resource registration is a mechanism for providing a yellow page registry for the discovery, manipulation, and integration of essential resources in the system, such as sensors and actuators. For pervasive applications, automatic discovery and the binding of system resources can greatly facilitate the system services and ease the effort of configuring, administering, and operating complex system resources.

Resource control is the management of various sensor and actuator resources in the system. Once resources are registered with the system, the middleware establishes control over these sensors and actuators for obtaining their states and for issuing control commands to them. Additionally, the middleware connects different devices so that they can communicate with each other via messages. The middleware also schedules resource usage throughout the system and potentially supports a large number of users.

Finally, resource abstraction allows other components in the middleware or application to use system resources, especially when users or devices move in that space.

2.2.2 User Tracking

Pervasive applications need to adapt their behavior to moving users. In order to do this, pervasive applications need an efficient user tracking method that can identify user positions and trails. Previous technologies for locating users' positions can be categorized into *indoor* and *outdoor* techniques.

Typically, for an outdoor environment, a GPS is used for obtaining position information, where that position information is calculated by measuring the distance between the GPS receiver and three or more GPS satellites. With the popularity of mobile phones, a similar location sensing technique has been developed by using Global System for Mobile communication (GSM) signals. Finally, video surveillance often is used in many cities to cover selected areas that typically need to be watched by humans or monitored by sophisticated computer vision software.

Indoor location sensing uses many wireless technologies, such as infrared [11], Bluetooth [12], and ultrasonic [13]. For instance, an active badge [11] attached to a person transmits infrared signals every few seconds that will be received by a network of sensors deployed in a building. Vision-based schemes capture videos to locate objects by vision recognition techniques and have achieved widespread usage. However, vision-based schemes [14] suffer from the line-of-sight problems because lights can be easily blocked by objects. With the availability of low-cost, general purpose RFID, this technology recently has been widely used in many applications. A typical RFID-based application attaches RFID tags to targeted objects beforehand. Then, either RFID readers or targeted objects move in space. When the tagged objects are within the accessible range of RFID readers, the information stored in the tags is emitted and received by readers. Thus, the reader knows the objects are within a nearby range.

2.2.3 Context Management

Context management includes *context acquisition*, *context representation*, and *context reasoning*. Context refers to the high-level information of an entity, which can be a person, a place, or an object that is relevant to the interaction between a user and pervasive applications. Thus, contexts can be classified into different categories, such as physical contexts of light, noise, and temperature, and user contexts of identities, social relationships, and activities.

Context acquisition is the process of collecting low-level information from various sensing sources. Such information is often captured via sensing devices. For instance, special sensors can gather information on temperature, magnetism, pressure, light, sound, and other chemical or mechanical properties. Actuators can collect information on motion and orientation. More complex sensors can capture images on videos of a targeted object. Finally, software sensors can report the I/O usage and network statistics on computers, pads, and smartphones.

Context representation refers to a formal description of the semantics of a context. For efficient processing of collected context information, we need to define structures of context information. Over the past decades, many different types of context representations have been developed, such as key-value pairs, logic-based models, object-oriented models, markup language models, and ontology-based models.

Context reasoning is the process of inferring high-level implicit context information from low-level explicit context information. Generally speaking, there are two types of context reasoning, *exact context reasoning* and *fuzzy context reasoning*. Exact context reasoning has been developed based on a Bayesian network [15], logic [16], cases [17], and ontology [18]. Fuzzy context reasoning includes methods based on evidence theory [19] and fuzzy logic [15].

2.2.4 Service Management

The middleware is typically structured as a number of services for a good reason. Services provide a much higher level of abstractions than those from the infrastructure layer. With carefully constructed services, pervasive applications easily can be built by directly invoking different services or service compositions. As a result, the middleware often supports mechanisms for efficient service management, including *service registration and discovery*, and *service composition*.

A pervasive environment is dynamic and heterogeneous in nature. As a result, different service registration strategies exist. Many systems in wired and infrastructure-based environments adopt service directories for maintaining service information and for processing queries and updates; these directories can be organized either as flat or a hierarchical trees. For pervasive environments, directories are dynamically placed on mobile nodes with better processing power, memory size, battery life, and node coverage. Another strategy is directory-less and uses broadcasting for disseminating service information. For mobile ad hoc networks (MANET), a distributed hash table (DHT)-based P2P overlay is often used for managing services [20].

Service composition is a way of combining elementary services in order to construct high-level services. In order to achieve service composition, service providers must first describe their services in a standard format. In the meantime, service consumers also need to specify their requests following some standard. Then, the middleware tries to fulfill the requested service by composing the provided services.

In the literature, there are mainly three different schemes for constructing a service composition plan. In the first scheme—from artificial intelligence [21,22]—atomic services are composed into a composite by treating each atomic service as an operator, and a planning algorithm links these operators to generate a plan. The second scheme uses the idea of work flow to describe how various services interact with one another [23]. Finally, a history-based composition scheme uses data mining to find previous usage patterns [24].

2.2.5 Data Management

Data management provides permanent data storage for pervasive applications. For pervasive applications, there are many mobile devices that are connected in a highly dynamic and ad hoc manner. As a result, data management must consider many issues, such as frequent change of data sources in terms of location and time, constant network disconnection among devices, and inconsistency of data. To make data management more challenging, existing data storage schemes developed for WSN cannot be directly applied for pervasive computing for several reasons.

First, pervasive applications have many different types of data sources, and storing a large amount of heterogeneous data is a nontrivial task. In addition, because many data sources exist, it is often necessary to resolve redundancy and inconsistency among collected data. The spatial and temporal properties of the data, as well as different data formats, make managing context data from different sources difficult.

Second, sensing data in a pervasive environment changes frequently, which can result in cascading updates among related data items. This is because a data item can share multiple contextual relationships (e.g., spatial, temporal, and social) with other items. Changing a single item causes updates of related items, which in turn results in further updates. Considering the unreliable communication layer, updating data items can be a challenging task.

Finally, both data producers and data consumers in a pervasive environment are more dynamic than in a WSN. They can join and leave the system at will, whereas sensor nodes in a WSN are often fixed and of the same type. Thus, pervasive applications have to work with unreliable and varied sensors.

The key function of data management is to provide seamless data access for pervasive applications. At the same time, certain Quality of Service (QoS) of data access should be provided in terms of latency, availability, and consistency. More specifically, data management not only stores a large amount of heterogeneous data but also efficiently maintains a contextual relationship among data objects.

2.2.6 Security Management

Security and privacy are important issues for pervasive applications because of the exposure of plenty of personal information, The traditional security problems include confidentiality, integrity, and availability. Confidentiality ensures that the users' data are not exposed to unauthorized users. Integrity maintains data without unauthorized alterations. Availability denotes the data are accessible when required. Pervasive applications introduce some more challenging security issues due to an extremely volatile nature—devices can join and leave arbitrarily, and communication links are often changing. In practice, systems often need to strike the right balance between security and privacy requirements and application requirements. For instance, many pervasive applications require users' location information to provide location-based services, such as finding a nearby ATM machine. However, disclosing the location information may raise serious privacy concerns.

Common security mechanisms, such as access control and user authentication, are required in the ubiquitous environment. Indeed, many projects, such as Gaia [5], use access control for security purposes. To protect users' privacy, *k-anonymity* has been used for many location-based services [25,26]. The idea is to use a location anonymizer to enlarge the queried regions to cover many other users. As a result, one user's real location is hard to distinguish from other users. In recent years, new security mechanisms such as voice recognition and facial recognition have appeared for enhanced user experiences.

Security is related to all other components in the middleware layer. For instance, user tracking is related to location privacy of users, and security mechanisms need to be utilized for service or resource discovery and registration. Otherwise, an active or passive intruder may gain unauthorized service, and there is a possibility that the intruder may even corrupt the system. To solve this problem, the Ninja project [27] proposes secure identification of services using a secure service discovery service (SSDS), where certificate authority (CA) is used. Other service discovery projects employ an encryption mechanism where a simplified version of public key infrastructure (PKI) is used [28].

2.3 Pervasive Computing Environments

2.3.1 Smart Car Space

Unlike previous smart spaces such as smart homes [29,30], the aware room [31], and the smart museum [32], the car space is highly mobile. The software system should be aware of this high mobility of vehicles. In addition, the vehicle space requires frequent information exchanges with the outside environment. For example, the vehicle needs local traffic information in order to adaptively choose different routes.

Figure 2.2 illustrates an implementation of smart car space [1]. The vehicle is installed with a touch screen, a GPS receiver, a video camera, and a wireless router. The GPS device enables position identification for the vehicle, and the video camera captures the driver image that is recognized by facial recognition. The wireless router provides an in-car network infrastructure to connect different components within the vehicle. As a result, when the user is recognized by the system, the smart car space can automatically trigger personalized services, such as enabling the entertainment system in the car to play the user's favorite music.

Figure 2.2   An illustration of smart car space.

A typical scenario in the smart car space is autonavigation with explicit environment feedbacks. The user in the car tells the system his destination via voice input. Then the system will process the voice input and select an optimal route. During driving, the system continuously monitors and reports the vehicle's status, such as regarding safety issues. For instance, the control system can detect that the road condition is bad and that the user is tired; then the user will be warned to focus on his driving for safety reasons. Additionally, different music may be played for the user according to the car speed and other contexts. For instance, the volume of the music will be tuned down when the user has an incoming phone call.

ScudWare [1] is a semantic and adaptive middleware platform for smart car space, where entities can intelligently adjust themselves adaptively to provide high-quality services for a changing environment. The features of ScudWare consist of:

- *Autonomy.* ScudWare dynamically monitors the environment for applications to interact autonomously.
- *Adaptability.* For changes in the inner vehicle running status and in the outside environment context, ScudWare can adaptively adjust the composition of its inner components and function accordingly.
- *Scalability.* ScudWare is a component-based middleware, where components can be dynamically added, replaced, and removed in a scalable fashion.
- *Semantic integration.* Semantic information is used to smoothly manage presentations and interactions of all entities in the smart car space for better collaboration among them.

The distinguishing feature of ScudWare is the integration of the ontology and the Ontology Web Language (OWL) language to support semantic context awareness. The applications in the smart car space are decomposed into many task units, and each task unit is serviced by a number of semantic virtual agents (SVAs). SVA is an abstract function union that comprises a number of meta-objects. Each meta-object denotes a kind of service comprising several components.

To manage context in the smart car space, a semantic context management service (SCMS) is responsible for context acquisition, representation, and fusion. A SCMS uses the ontology technique and the Web ontology language (OWL) to represent the sharing and reasoning of contexts.

For runtime adaptability and scalability, ScudWare uses an adaptive component management service (ACMS). An ACMS has two roles: (1) managing the component package, assembly, deployment, and allocation at design time and (2) managing component migration, replacement, updating, and variation at runtime. An ACMS allocates and reallocates components in the system, monitors component lifetime, and conducts the QoS of component execution.

ScudWare has been implemented on a Chinese Chery QQ car with four hardware parts: a central control unit, an engine control unit (ECU), smart sensors, and smart actuators. These parts communicate with each other through the control area net (CAN) – based network layer (CBL). With ScudWare middleware running on these hardware components, the smart car space offers a mobile music system and a smart navigation system.

2.3.2 Intelligent Campus

Intelligent campus (iCampus) is a pervasive computing environment providing minimal but flexible system support for people's daily activities on campus, as shown in Figure 2.3.

It provides a set of functionalities (e.g., resource management, user management, context awareness, message management, service management, and data management) and several typical campus applications, including

- *Map service.* It guides users to navigate on campus by providing detailed directions and some introduction related to the user's location.
- *Photo sharing service.* It can facilitate the sharing of photos and the construction of a social network.
- *Search service.* This school resource is valuable to both students and faculty; iCampus can search users' interested targets according to user preference.

To provide context-aware and adaptive services, iShadow is designed and developed based on middleware architecture. It exhibits the following important features.

- *Lightweight.* iCampus is aware of context and tracks users anytime, anywhere. For this purpose, iCampus consists of lightweight modules.
- *Scalability.* Based on a distributed resource discovery mechanism, iCampus can be easily extended to large-scale networks.

For more details, please refer to Chapter 8.

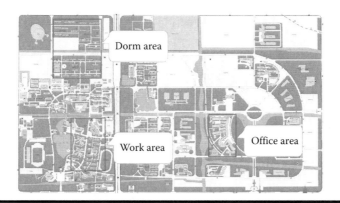

Figure 2.3   An illustration of iCampus.

Further Readings

Project Aura
http://www.cs.cmu.edu/~./aura/
Project Aura is a large-scale system demonstrating the concept of a *personal information aura* that spans wearable, handheld, desktop, and infrastructure computers. Aura's goal is to provide each user with an invisible halo of computing and information services that persists regardless of location.

Project Gaia
http://gaia.cs.uiuc.edu/
Project Gaia is a middleware operating system that manages the resources contained in an *active space* where physical spaces become interactive with users.

One.World
http://cs.nyu.edu/rgrimm/one.world/
One.World is a system architecture for pervasive computing that provides an integrated and comprehensive framework for building pervasive applications.

Middleware for Pervasive Computing: A Survey
By Vaskar Raychouhury et al., in *Pervasive and Mobile Computing*, pp. 177–200, 2013.
This survey paper gives a comprehensive illustration of existing pervasive middleware. An in-depth analysis of services provided by middleware systems is carried out, and some open research issues are discussed.

References

1. Z. Wu, Q. Wu, H. Cheng, G. Pan, M. Zhao, and J. Sun, ScudWare: A Semantic and Adaptive Middleware Platform for Smart Vehicle Space, *IEEE Transactions on Intelligent Transportation Systems*, vol. 8, no. 1, pp. 121–132, 2007.
2. Nike, FuelBand, 2014. [Online]. Available from: http://www.nike.com/us/en_us/c/nikeplus-fuelband, accessed on May 13, 2014.
3. Jawbone, Up. [Online]. Available from: https://jawbone.com/up, accessed on May 13, 2014.
4. Sony, SmartWatch, 2014. [Online]. Available from: http://www.sonymobile.com/us/products/accessories/smartwatch/, accessed on May 13, 2014.
5. M. Román, C. Hess, R. Cerqueira, A. Ranganathan, R. H. Campbell, and K. Nahrstedt, Gaia: A Middleware Platform for Active Spaces, *IEEE Pervasive Computing*, vol. 1, no. 4, pp. 74–83, 2002.
6. D. Garlan, D. P. Siewiorek, A. Smailagic, and P. Steenkiste, Project Aura: Toward Distraction-Free Pervasive Computing, *IEEE Pervasive Computing*, vol. 1, pp. 22–31, 2002.
7. M. Kumar, B. A. Shirazi, S. K. Das, B. Y. Sung, D. Levine, and M. Singhal, PICO: A Middleware Framework for Pervasive Computing, *IEEE Pervasive Computing*, vol. 2, pp. 72–79, 2003.

8. P. Veríssimo, V. Cahill, A. Casimiro, K. Cheverst, A. Friday, and J. Kaiser, CORTEX: Towards Supporting Autonomous and Cooperating Sentient Entities, in *Proceedings of the European Wireless*, 2002.

9. S. Kabadayi and C. Julien, A Local Data Abstraction and Communication Paradigm for Pervasive Computing, in *Proceedings of Fifth Annual IEEE International Conference on Pervasive Computing and Communications, PerCom 2007*, 2006, pp. 57–66.

10. J. Soldatos, L. Polymenakos, A. Pnevmatikakis, F. Talantzis, K. Stamatis, and M. Carras, Perceptual Interfaces and Distributed Agents Supporting Ubiquitous Computing Services, in *Proceedings of the Eurescom Summit 2005*, 2005, pp. 43–50.

11. R. Want, A. Hopper, V. Falcão, and J. Gibbons, The active badge location system, *ACM Transactions on Information Systems*, vol. 10, pp. 91–102, 1992.

12. F. D. Raffaele Bruno, Design and Analysis of a Bluetooth-Based Indoor Localization System, in *Proceedings of the 8th International Conference in Personal Wireless Communications*, 2003, pp. 711–725.

13. N. B. Priyantha, A. Chakraborty, and H. Balakrishnan, The Cricket Location-Support System, in *Proceedings of the 6th Annual International Conference on Mobile Computing and Networking—MobiCom '00*, 2000, pp. 32–43.

14. W. Hu, T. Tan, L. Wang, and S. Maybank, A Survey on Visual Surveillance of Object Motion and Behaviors, *IEEE Transactions on Systems, Man, and Cybernetics, Part C: Applications and Reviews*, vol. 34, no. 3, pp. 334–352, 2004.

15. A. Ranganathan, J. Al-Muhtadi, and R. H. H. Campbell, Reasoning about Uncertain Contexts in Pervasive Computing Environments, *IEEE Pervasive Computing*, vol. 3, no. 2, pp. 62–70, 2004.

16. A. Ranganathan and R. H. Campbell, An Infrastructure for Context-Awareness Based on First Order Logic, *Personal and Ubiquitous Computing*, vol. 7, no. 6, pp. 353–364, 2003.

17. D. Zhang, J. Cao, J. Zhou, and M. Guo, Extended Dempster-Shafer Theory in Context Reasoning for Ubiquitous Computing Environments, in *International Conference on Computational Science and Engineering*, 2009, pp. 205–212.

18. J. A. F. Nieto, M. E. B. Gutiérrez, and B. P. Lancho, Developing Home Care Intelligent Environments: From Theory to Practice, in *7th International Conference on Practical Applications of Agents and Multi-Agent Systems (PAAMS 2009)*, 2009, pp. 11–19.

19. D. Zhang, M. Guo, J. Zhou, D. Kang, and J. Cao, Context Reasoning Using Extended Evidence Theory in Pervasive Computing Environments, *Future Generation Computing Systems*, vol. 26, no. 2, pp. 207–216, 2010.

20. T. D. Hodes, S. E. Czerwinski, B. Y. Zhao, A. D. Joseph, and R. H. Katz, An Architecture for Secure Wide-Area Service Discovery, *Wireless Networks*, vol. 8, pp. 213–230, 2002.

21. M. Nidd, Service Discovery in DEAPspace, *IEEE Personal Communications*, vol. 8, pp. 39–45, 2001.

22. F. Sailhan and V. Issarny, Scalable Service Discovery for MANET, in *Proceedings of the Third IEEE International Conference on Pervasive Computing and Communications*, 2005, pp. 235–244.

23. M. Kim, M. Kumar, and B. Shirazi, Service Discovery Using Volunteer Nodes in Heterogeneous Pervasive Computing Environments, *Pervasive and Mobile Computing*, vol. 2, no. 3, pp. 313–343, 2006.

24. C. Lee and A. Helal, A Multi-Tier Ubiquitous Service Discovery Protocol for Mobile Clients, in *Proceedings of the 2003 International Symposium on Performance Evaluation of Computer and Telecommunication Systems (SPECTS)*, 2003.

25. B. Niu, Q. Li, X. Zhu, G. Cao, and H. Li, Achieving k-Anonymity in Privacy-Aware Location-Based Services, in *INFOCOM*, 2014, pp. 754–762.

26. L. Sweeney, k-Anonymity: A Model for Protecting Privacy, *International Journal of Uncertainty, Fuzziness and Knowledge-Based Systems*, vol. 10, no. 5, pp. 557–570, 2002.

27. J. R. von Behren, E. A. Brewer, N. Borisov, M. Chen, M. Welsh, J. MacDonald, J. Lau, S. Gribble, and D. Culler, Ninja: A Framework for Network Services, in *USENIX Annual Technical Conference*, 2002.

28. J. Undercoffer, F. Perich, A. Cedilnik, L. Kagal, and A. Joshi, A Secure Infrastructure for Service Discovery and Access in Pervasive Computing, *Mobile Networks and Applications*, vol. 8, no. 2, pp. 113–125, 2003.

29. Smart Room, Media Lab, Massachusetts Institute of Technology. [Online]. Available from: http://vismod.www.media.mit.edu/vismod/demos/smartroom/.

30. Kids Room, Media Lab, Massachusetts Institute of Technology. [Online]. Available from: http://vismod.www.media.mit.edu/vismod/demos/kidsroom/.

31. J. A. Kientz, S. N. Patel, B. Jones, E. Price, E. D. Mynatt, and G. D. Abowd, The Georgia Tech Aware Home, in *Extended Abstracts on Human Factors in Computing Systems*, 2008, pp. 3675–3680.

32. M. Fleck, M. Frid, T. Kindberg, E. O'Brien-Strain, R. Rajani, and M. Spasojevic, From Informing to Remembering: Ubiquitous Systems in Interactive Museums, *IEEE Pervasive Computing*, vol. 1, no. 2, pp. 13–21, 2002.

Chapter 3

Context Collection, User Tracking, and Context Reasoning

Context refers either to aspects of the physical world or to conditions and activities in the virtual world. As a kernel technique, context awareness plays a key role in pervasive computing. By context awareness, pervasive systems can sense their environments and adapt their behavior accordingly.

This chapter introduces how to collect pervasive context through wireless sensor networks (WSNs) and then presents a user tracking scheme in Section 3.2. Finally, Section 3.3 discusses various reasoning models and techniques, together with performance evaluation.

3.1 Context Collection and Wireless Sensor Networks

Context can be any information that characterizes the situation of various entities in pervasive environments, including persons, places, locations, or objects relevant to the interaction between users and applications. Researchers have defined pervasive context from different points of views. For example, Dey and colleagues categorized context as location, identity, activity, and time [1]; yet Kaltz et al. [2] identified the categories of user/role, process/task, location, time, and device. They also emphasized, for these classical modalities, that any optimal categorization greatly depends on the application domain and usage.

In fact, context is an open concept because it is not limited by one's imagination. Any system that exploits available context information needs to define the scope of the corresponding context. It is not always possible to enumerate a limited set of contexts that match a real-world context in all context-aware applications. So, we will first discuss characteristics and general categorization of context in pervasive systems.

3.1.1 Context Category

Pervasive computing needs to be based on the user's life, and it then adds the selected device to the user. Next, context awareness follows steps to provide services intelligently, according to the context information. Situational awareness is the starting point of pervasive computing; it turns off the computer operations in our daily lives. To collect context data efficiently, we divide context in pervasive systems into two models. The first is direct context (i.e., low-level context), which is a relatively generic model. It can be obtained from sensors, digital devices, and so on. The second is indirect context (i.e., the highest context level), which can be derived from direct context.

Currently, pervasive computing is advancing along with the rapid development of computers because pervasive systems have to provide users with transparent services through context awareness and further adaptivity to various context. More specifically, context awareness should enable applications to extend capabilities of devices as well as their resources by using those available in the local environment; pervasive computing devices are, in general, resource constrained. Next, context awareness should allow users to take advantage of unexpected opportunities. For example, users could be informed that they will be passing a gas station with prices significantly less expensive than those at their usual station. Finally, context awareness will allow information to be prefetched in anticipation of future contexts, thereby providing better performance and availability. For instance, when applications access calendar and location information, they can prefetch maps, restaurant information, and local weather/traffic conditions.

Context in pervasive systems exhibits certain characteristics. The first is *dynamics*. Although some context dimensions are static (such as a username), most of the context dimensions are highly dynamic (such as the location of a user). Furthermore, some context dimensions change more frequently than others. Also, some context has an evolving nature, that is, it is continuously changing over time. Next, context is *interdependent*. Context dimensions are somehow interrelated. For instance, there are different kinds of correlations between an individual's home and his job. What is more, context may be *fuzzy*. Some context is imprecise and incomplete. It is a known fact that sensors are not 100% accurate. Besides, multiple sensors might provide different values for the same context.

Classifying context is important for managing context quality as well as understanding the context and application development. In Figure 3.1, we categorize

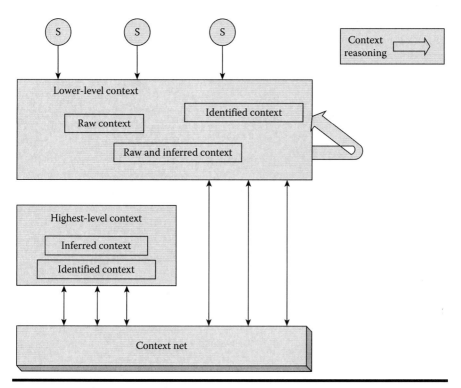

Figure 3.1   Different levels of context.

various context based on their characteristics, according to different classification principles.

From the collection point of view, context can be categorized as *direct context* and *indirect context*. Direct context refers to gathered information that does not involve any extra processing. If the information is gathered implicitly by means of sensors, it is called *sensed context*. If the information is gathered explicitly, it is called *defined context*. Indirect context refers to one inferring information from direct context.

From the application point of view, context falls into *low-level context* and *high-level context*. Low-level context information is usually gathered from sensors or from application logs, which can be considered environmental atomic facts. High-level context information is derived from low-level context information.

Finally, context information can be categorized into two categories from a temporal point of view: *static context* and *dynamic context*. Static context does not change with time (such as a person's race). Dynamic context keeps changing in different frequencies depending on the context dimension (such as a person's location or age). This implies that, for a dynamic context dimension, various values might be available. Hence, management of context information's temporal character is crucial either in the sense of historical context or in the sense of the validity of available contextual information.

3.1.2 Context Collection Framework

The goal for all context data collection is to capture quality evidence that then translates to rich data analysis and that allows the building of a convincing and credible answer to questions that have been posed. As discussed earlier, context is not simply the state of a fixed set of interaction resources, it is also a process of interacting with an ever-changing environment composed of reconfigurable, migratory, distributed, and multiscale resources. So, dynamic context has to collect during the execution process of pervasive applications, providing context-aware pervasive computing with adaptive capacity for dynamically changing environments. The data collection component of research is common to all fields of study including physical and social sciences, humanities, and business.

WSNs can be used to collect the dynamic context in pervasive systems efficiently due to their flexibility, fault tolerance, high sensing, self-organization, fidelity, low cost, and rapid deployment [3,4]. However, the traditional flat architecture of WSNs, where all sensor nodes send their data to a single sink node by multiple hops, limits large-scale pervasive context collection [5–7].

Figure 3.2 shows an architecture that merges the advantages of mesh networks and WSNs through deploying multiple wireless mesh routers equipped with gateways in each sensor network. It can provide capacities to interconnect multiple homogeneous/heterogeneous sensor networks; to improve the scalability, robustness, and data throughput of sensor networks; and to support the mobility of nodes.

Mesh routers in Figure 3.2 automatically interconnect to form a mesh network while they are connected with the Internet through powerful base stations. In such an architecture, there are three kinds of networks on three logical layers:

- *Wireless sensor network* for monitoring objects and reporting the objects' information (e.g., temperature and humidity)
- *Wireless mesh network* for transmitting sensed data long distance and in a reliable way
- *Internet* for users to remotely access sensed data

Accordingly, a wireless mesh sensor network (WMSN) is composed of three kinds of nodes: a *sensor node*, a *wireless mesh gateway* (WMG), and a *wireless mesh router* (WMR). In particular, base stations are used to support the mobility of WMGs and WMRs, and they connect the wireless mesh network with the Internet. Sensor nodes continuously or intermittently detect objects and then send data to the most appropriate WMG based on specific routing policies, which will be discussed in detail in the next section. WMGs work as sink nodes and gateways

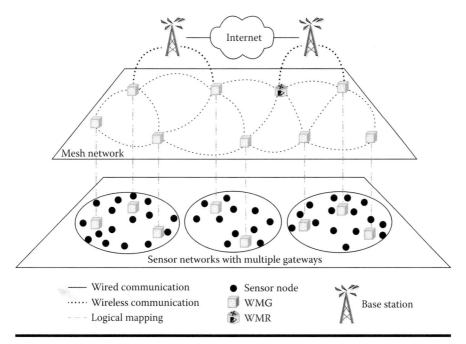

Figure 3.2   A scalable wireless mesh sensor network architecture.

for low-level WSNs as well as routers for middle-level wireless mesh networks. By comparison, WMRs only serve as routers for wireless mesh networks. WMGs and WMRs self-organize as the middle-level wireless mesh network for long-distance transmission and interconnection of sensor networks.

3.2 User Tracking

Location information is one of the most important contexts in pervasive comput-ing [8–10]. With location information, pervasive applications can sense users' positions and frequent routes, and they can provide location-based, customized service for different users.

In this section, we first discuss technologies for identifying user positions and then we describe a position identification for a mobile robot. Finally, we discuss our experience in managing urban traffic in the city of Shanghai.

3.2.1 Position Identification

Generally speaking, the technologies for locating user position can be categorized into two types, *indoor* and *outdoor* techniques. Typically, the outdoor environment uses global positioning system (GPS) sensors for obtaining position information.

For an indoor environment, wireless technologies, such as infrared [11], Bluetooth [12,13], and ultrasonic [14,15], have fostered a number of approaches in location sensing. The availability of low-cost, general purpose radio frequency identification (RFID) technology [16–19], has resulted in its being widely used. A typical RFID-based application attaches RFID tags to targeted objects beforehand. Then, either RFID readers or targeted objects move in space. When the tagged objects are within the accessible range of RFID readers, the information stored in the tags is emitted and received by the readers. Thus, the readers know when the objects are within a nearby range.

We have used RFID tags for tracking users' positions. In particular, we have designed a tag-free activity sensing (TASA) approach using RFID tag arrays for location sensing and route tracking [20]. TASA keeps the monitored objects from attaching RFID tags and is capable of sensing concurrent activity online. TASA deploys passive RFID tags into an array, captures the received signal strength indicator (RSSI) series for all tags all the time, and recovers trajectories by exploring variations in RSSI series. We have also used a similar idea for locating a moving object via readings from RFID tag arrays [16] but using mostly passive tags instead of active tags.

3.2.2 Mobile Robot Position Identification

With technology advancement, home robots are expected to assist humans in a number of different application scenarios. A key capability of these robots is being able to navigate in a known environment in a goal-directed and efficient manner, thereby requiring the robot to determine its own position and orientation.

We have designed a visual position calibration method using colored rectangle signboards for a mobile robot to locate its position and orientation [21]. Figure 3.3 shows our solution for indoor robot navigation. In that environment, the space is divided into many subareas. Between two subareas, there is a relay point in the center of each path. Near each relay point, we set a specially designed signboard for robot position calibration. Every time the robot passes a relay point, the robot will have a chance to calibrate its cumulative position errors. In order to position the signboard in the center of the image, sound beacons were used to control the robot's head to make it turn directly toward the landmarks [22].

The signboards can use distinguishing colors (e.g., red) so that they can be easily detected from the background, simplifying the region extraction. The rectangular shape is convenient for using the vanishing point method to calculate the relative direction and distance between the robot and the signboards. Using that perspective, two parallel lines projected in a plane surface will intersect at a certain point—called a *vanishing point*. Vanishing points in a perspective image have been used for reconstructing objects in three-dimensional space with one or more images from different angles [23–25]. The vanishing method can also calculate the relative direction of an object to a perspective point (i.e., the center position of the camera).

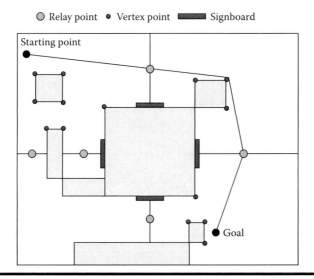

Figure 3.3 **An environmental map with specially designed signboards for position calibration.**

A Canny algorithm is designed to detect the edge of the color area to calculate the vanishing points in the image of the signboard [21]. The advantage of our approach is that it can obtain the relative direction and distance between the signboard and the robot with only a single image. Experimental results showed our method is effective; the errors were between about 1.5 degrees and 3 cm in an array position.

3.2.3 Intelligent Urban Traffic Management

We have developed an intelligent urban traffic management system that not only provides traffic guidance, such as a road status report (free or jammed), but also offers the best path prediction and remaining time estimation. Specifically, from 2006 to 2010, we installed GPS and wireless communication devices on more than 4000 taxis and 2000 buses as a part of the Shanghai Grid project [26]. Each vehicle collects GPS location data and reports the data via a wireless device to a central server. By monitoring a large number of public transportation vehicles, the system performs traffic surveillance and offers traffic information sharing on the Internet.

- *Real-time road status report.* The road status denotes the average traffic flux of a road. The bigger the traffic flux is for a road, the more crowded the road is and the slower the traffic moves. From the GPS data of vehicle location and speed information, we compute the average speed of vehicles on a road.

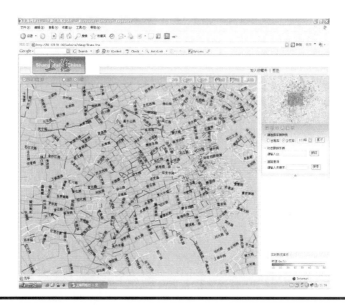

Figure 3.4   Real-time traffic conditions in Shanghai.

As an example, Figure 3.4 shows a snapshot of a real-time traffic condition in Shanghai, where a line represents a road and different colors indicate various traffic conditions: roads colored in black, light gray, and gray denote congestion, light congestion, and normal conditions, respectively. With this application, the traffic information is visualized for the user and is shown in public e-displays. Users can also access the information with Web browsers.

■ *The best path prediction.* The best path refers to a sequence of roads from one place to a specified destination, with a minimal cost dependent on different prediction policies. The best path between two specified places may change from time to time. For example, the best path based on the least time policy is different at different times due to changing road status.

The Best Path Prediction service finds the best path to enable people to travel in Shanghai more conveniently. Currently, this service supports the least time policy and the least fee policy. Figure 3.5 illustrates a recommended route that requires the least time to travel from Huaihai Road to Liyang Road, with a sequence of road names at the top left corner. Users can access this service using Web browsers or mobile phones. For a mobile phone user, the service returns a short message service (SMS) showing a sequence of road names.

■ *Remaining time estimation.* This application computes the estimated remaining time to a destination (for instance, determining how long before the next bus will arrive at a stop). This kind of question is answered considering real-time traffic information, road and intersection maps, and a vehicle's position and direction.

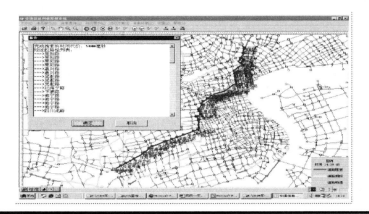

Figure 3.5   A recommended route with the shortest travel time.

3.3 Context Reasoning

In pervasive computing, context reasoning enables context awareness for ubiquitous applications. Specifically, context reasoning infers high-level implicit context information from the low-level explicit context information. Existing context reasoning techniques can be classified into two types—exact reasoning and inexact or fuzzy reasoning. In this section, we present an inexact reasoning method, the extended Dempster–Shafer theory for context reasoning (DSCR), in ubiquitous/pervasive computing environments.

Figure 3.6 illustrates our context-aware architecture, which is a hierarchical module consisting of context providers, a manager, and consumers. The context providers gather context data on the environments from sensors. Note that contexts may come from applications and that our context collection is capable of gathering contexts at intervals or on demand. The context manager preprocesses contexts and responds to requests from adaptors and services. After the context information is generated, the context manager will format it, remove inconsistencies in the preprocessing step, and then move into the reasoning step where hidden contexts are derived from present and historical contexts. A historical context database is introduced to store the past contexts because contexts that are updated frequently increase considerably in size. Finally, the context manager notifies adaptors and services and distributes contexts to them according to their specific privacy and security requirements and their resource constraints. Here, we implement a prototype to demonstrate our context-aware architecture. We employ TinyOS 2.0, Micaz, and Microsoft SQL servers to collect, preprocess, and store contexts. Our prototype currently supports devices with Bluetooth, Wi-Fi, and WLAN.

It is noteworthy to point out that DSCR is applied to context collection, reasoning, and storage in the context manager. This enables the context manager to handle imprecise and incomplete contexts.

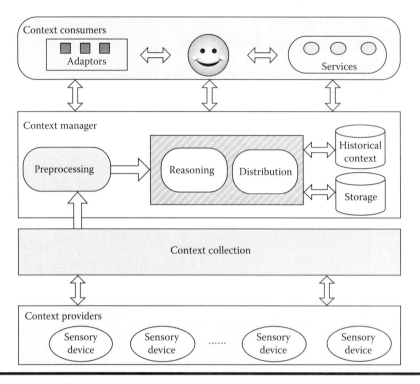

Figure 3.6 The architecture of context-aware pervasive applications. Context providers are responsible for collecting contextual information, which is preprocessed by the context manager in order to derive hidden contexts from present and historical contexts. Finally, the reasoning results are delivered to context consumers.

The rest of this section gives an overview of evidence theory and then discusses our DSCR model in detail. Finally, we discuss the performance of our model in comparison to previous approaches.

3.3.1 Evidence Theory

Evidence theory is a mathematical theory that constructs a coherent picture of reality through computing the probability of an event given evidence [27,28]. Evidence theory is based on the following concepts and principles:

■ *Frame of discernment.* Let τ be the frame of discernment, denoting a set of mutually exclusive and exhaustive hypotheses about problem domains. The set 2^τ is the power set of τ, denoting the set of all subsets of τ.

- *Mass.* Mass stands for a belief mapping from 2^τ to the interval between 0 and 1, represented as *m*. Let ø and *C* be the empty set and a subset of τ. The masses of the empty set and the sum of all the subsets in the power set are 0 and 1, respectively. Mass can be assigned to sets or intervals.
- *Belief and plausibility.* The belief of a hypothesis is the sum of the beliefs for that hypothesis that are its subsets. Conversely, the plausibility of a hypothesis is the sum of all the beliefs of sets that intersect with it. Belief and plausibility are defined as:

$$Bel(C)= \sum_{B|B\subseteq C} m(B), \quad Pls(C)= \sum_{B|B\cap C\neq\emptyset} m(B) \qquad (3.1)$$

where *B* is a subset of τ. *Bel* is the degree of belief to which the evidence supports *C*, constituted by the sum of the masses of all sets enclosed by it. *Pls* denotes the degree of belief to which the evidence fails to refute *C*, that is, the degree of belief to which it remains plausible (i.e., the possibility that the hypothesis could happen).

- *Dempster's rule.* In order to aggregate the evidence from multiple sources, Dempster's rule plays a significantly meaningful and interesting part. Let C_i be a subset of τ and $m_j(C_i)$ be a mass assignment for hypothesis C_i collected from the *j*th source. For subsets $\{C_1, C_2, ..., C_n\}$ and mass assignments $\{m_1, m_2, ..., m_n\}$, Dempster's rule is given as:

$$m\{C\}=(m_1 \oplus m_2 \oplus ... \oplus m_n)(C)=\frac{1}{K}\sum_{\tau=C} m_1(C_1)m_2(C_2)...m_n(C_n) \quad (3.2)$$

where τ equals to $C_1 \cap C_2 \cap ... \cap C_n$, and *K* is the normalizing constant that is defined as:

$$K=1-\sum_{\tau=\emptyset} m_1(C_1)m_2(C_2)...m_n(C_n) \qquad (3.3)$$

Equations 3.2 and 3.3 show that Dempster's rule combines evidence over the set of all evidence. Its complexity grows exponentially with increases in the amount of evidence. In Orponen's study [29], the complexity of Dempster's rule is proved as NP-complete. Meanwhile, once some probabilities of evidence are zero, Dempster's rule will get an incorrect inference that a less probable event is regarded as the most probable event that could happen. This phenomenon is called the Zadeh paradox.

3.3.2 DSCR Model

To facilitate understanding of our argument, we present a formal description and a model for DSCR. Let $S = \{s_1, s_2, ..., s_n\}$, $O = \{o_1, o_2, ..., o_n\}$, and $A = \{a_1, a_2, ..., a_n\}$ be the sets of sensors, objects, and activities, respectively. Let SO be simple objects monitored by sensors directly, DO be deduced objects obtained from SO, CO be composite objects, E be the set of evidence, and D be the discount rate, reflecting the reliability of sensor readings.

We model DSCR as sensors, objects, and activity layers. Sensors are deployed to monitor objects in the sensors layer. All objects are in the objects layer, where some objects are monitored by sensors. Activity is in the activities layer, which can be inferred according to the contexts of objects. In Figure 3.7, objects o_5 and o_6 are a deduced object and a composite object, respectively, and lines with arrows denote relationships among sensors, objects, and activities. Objects o_1, o_2, o_3, and o_4 are monitored by sensors s_1, s_2, s_3, and s_4, respectively. Object o_5 is deduced from o_1, and o_6 is made up of objects o_2 and o_3. Let us suppose a scenario where a sensor is deployed to monitor the door of a refrigerator where coffee, milk, and tea are stored. When the door is open, users makes drink by selecting coffee, milk, or tea, or their combinations—as simple a_1 or a_2, or complex activity a_3. In this case, the refrigerator is a type of object o_4; coffee, milk, and tea are each a type of object o_5; and their combinations are types of object o_6.

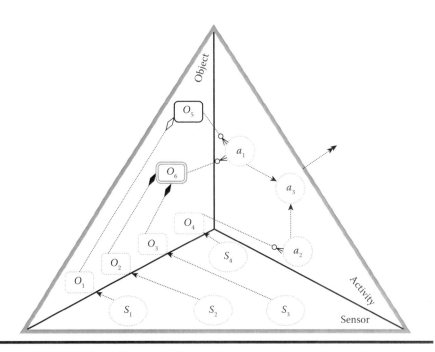

Figure 3.7   DSCR working model.

In Sections 3.3.3, 3.3.4, and 3.3.5, we describe how to apply DSCR to this scenario. The process of applying DSCR consists of (1) propagating evidence in the sensors layer, (2) propagating evidence in the objects layer, and (3) recognizing activity. In Section 3.3.6, we design a computation reduction strategy for DSCR.

3.3.3 Propagating Evidence in the Sensors Layer

Sensors are vulnerable and vary drastically with environments, which leads to unreliable sensor readings. Therefore, it is important to take into account the reliability of sensors in the evidence aggregation process. This reliability is controlled by the sensor discount rate, which is derived from the sensor reliability model [30,31]. In DSCR, the sensor discount rate is incorporated and is defined in Equation 3.4:

$$m^r(C) = \begin{cases} (1-r)m(C) & C \neq \tau \\ r + (1-r)m(\tau) & \text{otherwise} \end{cases} \tag{3.4}$$

where r is the sensor discount rate with its value between 0 and 1. When r is 0, the sensor is completely reliable; when r is 1, the sensor is totally unreliable.

In the beginning, sensors s_1, s_2, s_3, and s_4 are installed and working with discount rates set at 0.02, 0.02, 0.1, and 0.2, respectively. Evidence on sensor nodes is represented by masses as: $m_{s_1}(s_1) = 1$, $m_{s_2}(s_2) = 1$, $m_{s_3}(\neg s_3) = 1$, and $m_{s_4}(s_4) = 1$. Then, the discounted masses are calculated as:

$$
\begin{aligned}
m^r(s_1) &= 0.98; & m^r(s_1, \neg s_1) &= 0.98 \\
m^r(s_2) &= 0.98; & m^r(s_2, \neg s_2) &= 0.98 \\
m^r(\neg s_3) &= 0.9; & m^r(s_3, \neg s_3) &= 0.1 \\
m^r(s_4) &= 0.8; & m^r(s_4, \neg s_4) &= 0.2
\end{aligned}
$$

There is a relationship between sensors and their associated objects. For example, given the frames τ_a and τ_b of sensor s_a and its associated object o_b, the relationship between sensor s_a and object o_b denotes the evidence propagation from sensors to the objects layer. DSCR associates objects with sensors by maintaining a compatible relationship between them, which is defined as evidential mapping by Equation 3.5:

$$\Gamma : m(o_b) = f(s_a \rightarrow o_b) \tag{3.5}$$

where f is a mapping function, propagating the evidence from sensor s_a to object o_b. For objects o_1, o_2, o_3, and o_4 in Figure 3.7, their masses are computed as:

$$
\begin{aligned}
\Gamma : m(o_1) &= m^r(s_1) &= 0.98 \\
\Gamma : m(o_1, \neg o_1) &= m^r(s_1, \neg s_1) &= 0.02
\end{aligned}
$$

$$\Gamma : m(o_2) = m^r(s_2) = 0.98$$
$$\Gamma : m(o_2, \neg o_2) = m^r(s_2, \neg s_2) = 0.02$$
$$\Gamma : m(o_3) = m^r(s_3) = 0.9$$
$$\Gamma : m(o_3, \neg o_3) = m^r(s_3, \neg s_3) = 0.1$$
$$\Gamma : m(o_4) = m^r(s_4) = 0.8$$
$$\Gamma : m(o_4, \neg o_4) = m^r(s_4, \neg s_4) = 0.2$$

In this step, evidence is propagated to all the sensors and objects. With the help of the mapping function, the masses of sensors are transferred to the objects layer.

3.3.4 Propagating Evidence in the Objects Layer

After the masses of sensors are transferred, this step further propagates the evidence in the objects layer. Suppose we collect evidence from the observations illustrated in Table 3.1. Note that evidence $\{o_1\} \rightarrow \{o_5\}$ refers to a 0.9 confidence about object o_5 when we observe object o_1. We take simple object o_4, deduced object o_5, and composite object o_6 to explain this step.

■ *Calculating masses for the deduced objects: DO $\leftarrow$ E.* Although the deduced objects are not monitored directly by sensors, evidential mapping is capable of calculating their masses from their evidence. For instance, evidential mapping propagates masses from objects o_1 to o_5 as:

$$m(\{o_5\}) = m(\{o_1\}) * m(\{o_1\} \rightarrow \{o_5\}) = 0.882$$
$$m(\{o_5, \neg o_5\}) = m(\{o_1\}) * m(\{o_1\} \rightarrow \{o_5, \neg o_5\})$$
$$+ m(\{o_1, \neg o_1\}) * m(\{o_1, \neg o_1\} \rightarrow \{o_5, \neg o_5\}) = 0.118$$

Table 3.1 Evidential Mapping

Relationship		Evidence Value
$\{o_1\}$	$\rightarrow \{o_5\}$	0.9
$\{\neg o_1\}$	$\rightarrow \{\neg o_5\}$	1.0
$\{o_1\}$	$\rightarrow \{o_5, \neg o_5\}$	0.1
$\{o_1, \neg o_1\}$	$\rightarrow \{o_5, \neg o_5\}$	1.0
$\{o_4\}$	$\rightarrow \{a_2\}$	0.3
$\{\neg o_4\}$	$\rightarrow \{\neg a_2\}$	1.0
$\{o_4\}$	$\rightarrow \{a_2, \neg a_2\}$	0.7
$\{o_4, \neg o_4\}$	$\rightarrow \{a_2, \neg a_2\}$	1.0

■ *Calculating masses for the composite objects: CO ← E.* This step aims to propagate the evidence to the composite objects. This is also achieved by evidential mapping and is much more complex than that of the deduced objects. It includes two sub-steps: (1) calculating masses to the sets of objects that comprise the composite objects and (2) calculating masses of the composite objects. In the first sub-step, DSCR computes the masses of these sets using Equation 3.5:

$$\Gamma : m_1(o_2, o_3) = m(\{o_2\}) = 0.98$$
$$\Gamma : m_2(o_2, o_3) = m(\{o_3\}) = 0.9$$
$$\Gamma : m_1(\{o_2, o_3\}, \neg\{o_2, o_3\}) = m(\{o_2, \neg o_2\}) = 0.02$$
$$\Gamma : m_2(\{o_2, o_3\}, \neg\{o_2, o_3\}) = m(\{o_3, \neg o_3\}) = 0.1$$

Thus, we get different beliefs from two sources (i.e., sensors s_2 and s_3). How to generate the masses of the composite objects is an interesting process. Most existing approaches ignore the difference among the evidence from multiple sources; we use a weighted sum to compute the masses of composite objects, where the weight corresponds to the discount rate for the mass.

■ *Generating an activity candidate set from objects: A ← O.* So far, we can generate an activity candidate set in which users may undertake one or more activities at a time. *Even though knowing precisely what the users want is impossible, observations show that most routine tasks are predictable.* In fact, an object–activity mapping table is collected in DSCR that conforms closely to the reality of how users work. This mapping table varies with each scenario, and data can be collected from observations. For instance, in Figure 3.7, activity a_1 obtains the masses propagated by objects o_5 and o_6 according to the object–activity mapping table. Then its mass is calculated as:

$$m_1(\{a_1\}) = m(\{o_5\}) = 0.882$$
$$m_2(\{a_1\}) = m(\{o_6\}) = 0.948$$
$$m_1(\{a_1, \neg a_1\}) = m(\{o_5, \neg o_5\}) = 0.118$$
$$m_2(\{a_1, \neg a_1\}) = m(\{o_6, \neg o_6\}) = 0.052$$

For the same reason, the mass of activity a_2 needs to be calculated. At this point, DSCR has generated an activity candidate set and is ready for user activity recognition in the next step.

3.3.5 Recognizing User Activity

This step focuses on recognizing user activity from the activity candidate set. For example, activity a_1 is chosen because of its bigger mass than that of other activities. Observations in previous steps reveal that the original accuracy of sensors has a significant impact on the final results. Two methods are available to avoid the

Table 3.2 Example of an Activity Candidate Set

Activity	m_1	m_2	$\underline{m}$ - Final Result
$\{a_1\}$	0.40	0.20	?
$\{a_2\}$	0.30	0.20	?
$\{a_3\}$	0.20	0.30	?
$\{a_1, a_2\}$	0.05	0.15	?
$\{a_2, a_3\}$	0.04	0.10	?
$\Theta = \{a_1, a_2, a_3\}$	0.01	0.05	?

influence of sensor reliability. One method is to use powerful sensors to improve accuracy, but this increases cost and prevents ubiquitous computing. The other method is to use multiple sensors to monitor the same objects or to sample the same context several times. In most cases, the latter method is preferred for ubiquitous applications because of lower cost.

To clearly describe the mechanism of activity recognition, we extend the example used earlier. Suppose that we collect the evidence for the same scenario and process it by the steps mentioned before (see Table 3.2). According to Equation 3.3, the normalizing constant K is calculated to be 0.477. Then, using Equation 3.2, the combination mass of activity a_1 is computed to be 0.3606.

In the same way, we can calculate the combined masses $\underline{m}\{a_2\}$, $\underline{m}\{a_3\}$, $\underline{m}\{a_1, a_2\}$, $\underline{m}\{a_2, a_3\}$, and $\underline{m}\{\tau\}$ as 0.3795, 0.2201, 0.0241, 0.0147, and 0.0010, respectively. After all the combination masses have been obtained, applications will compute the beliefs using Equation 3.1 and then will recognize the activity. Correspondingly, activity a_2 in Table 3.2 is selected as the prediction. Up to now, all the steps of applying DSCR into ubiquitous computing environments have been described. However, the problem is still not solved—the intensive computation that arises from calculating the normalizing constant and combined masses [28,29].

3.3.6 Evidence Selection Strategy

In order to solve this problem, we present a strategy—that of evidence selection for Dempster's rule. The calculation for the normalizing constant K in Dempster's rule is time-consuming because it requires traversing all the evidence and masses.

We find that the observation illustrated as Lemma 3.1 can reduce the operations for the normalizing constant to $|\tau|$. Note that Dempster's rule, which uses Equation 3.3 to calculate the normalizing constant, is appropriate for calculating the combined mass for any one hypothesis without calculating all the combined masses.

Lemma 3.1

The normalizing constant K can be computed from all the masses that are to be normalized.

The other computation overhead of DSCR emerges from the calculation of the combined masses. Suppose we calculate the combined mass for evidence C by calculating its unnormalized mass and then normalizing it. The total number of operations required is proportional to $\log(|C|)+\log(|\tau|)$ for ordered lists but is exponential in $|\tau|$ for n-dimensional array representation. In view of the fact that the decisions made by ubiquitous applications in most cases rely on the most related contexts, DSCR employs a process of selecting evidence. This process consists of two sub-steps.

First, DSCR defines the importance of evidence by t_i that is given as:

$$t_i = \sum_{j=1}^{n} w_j * m_j, \tag{3.6}$$

where j is the number of masses and w_j is the weight of mass m_j for evidence i. Consider that masses from multiple sources have different impacts on the decision making for ubiquitous applications. DSCR is capable of tuning w_j to emphasize the most important masses.

Second, DSCR selects the most related evidence by k-l strategy. Let n be the number of evidence, k be the minimum number of evidence to be kept, l be the maximum number of evidence to be kept, η be the threshold on the sum of masses of selected evidence, Σ be the sum of masses of selected evidence, ϑ be the number of selected evidence, and LSP be the least small probability event. Lines 5 to 17 in Algorithm 3.1 are k-l selection strategy for evidence selection. When l is equal to k, the selection strategy turns into top-K selection strategy. Note that the step of randomized quicksort in k-l strategy is just marking the order of evidence according to the t_i and does not swap evidence orders during the sort process.

For k-l evidence selection strategy, the time complexity is $O(n)$, where n is the amount of evidence. The space complexity is also $O(n)$. For the conflict resolution strategy, the time complexity is $O(\vartheta)$, where ϑ is the amount of evidence used in Dempster's rule. DSCR is scalable even for a large amount of evidence because ϑ is much less than n.

3.3.7 Performance

We compare the performance of DSCR with the typical Bayesian network approach [32–34], D–S theory, and with the Yager [35] and Smets [36] approaches. Note that both rule-based and logic-based approaches are designed for exact reasoning, which cannot work well with imprecise and incomplete context data.

Algorithm 3.1   The Optimized Selection Algorithm

1:	**input:** e: evidence set
2:	**output:** e^t: selected $k - l$ evidence
3:	a_i: the prediction result
4:	
5:	// $k - l$ selection strategy
6:	// calculate the weighted sum of masses
7:	**for** $i \leftarrow 1, n$ **do**
8:	compute t_i
9:	// rst: result of all t_i
10:	$rst \leftarrow t_i$
11:	**end for**
12:	
13:	randomized quicksort e based on t
14:	// select top k evidence
15:	$e^t \leftarrow e(k)$
16:	**while** $(\Sigma > \eta \ \&\& \ \vartheta < l)$ **do**
17:	$e' \leftarrow$ the evidence with the next highest mass
18:	**end while**

Ten user activities are extracted from the MIT *Houser_n* project, which aimed to facilitate user real life by continuous monitoring and recording all activities of users [37]. The *Houser_n* project provided the videos, logs, datasets, and nitty-gritty (true activity). In the experiments, we selected singleton and complex activity as test activities, together with associated information, such as their frequency during a sampled month and the surrounding contexts when they happened. Due to a lack of information related to the discount rate of all sensors in the dataset, we assume all sensors work properly with zero discount rate. Finally, we run the experiments nine times with context data from a different month each time.

Figure 3.8 shows the overall performance of DSCR. The x-axis is the i-th experiment. DSCR outperforms the other approaches by about 28.65% in recognition rate. D–S theory achieves a better result than the Bayesian network. This is because the evidence used in these experiments does not contain much highly conflicting evidence. The Yager and Smets approaches alleviate the influence of conflicting

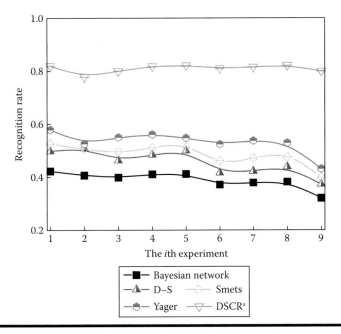

Figure 3.8   Recognition rate of recognition user activity. The higher the recognition rate, the better the performance.

evidence and achieve the higher recognition rate. This implies that some evidence is highly conflicting in our test set. Contrary to this, the Bayesian network cannot achieve a high recognition rate.

Table 3.2 explains why the Bayesian network fails to handle such evidence. Let α, β, γ, η, ξ, and τ denote six hypotheses a_1, a_2, a_3, $\{a_1, a_2\}$, $\{a_2, a_3\}$ and $\{a_1, a_2, a_3\}$, respectively. First, α relates to η and τ, while η, ξ, and τ correlate to each other. They are not exclusive and exhaustive (e.g., $\{a_1, a_3\}$ is missing). Second, the causal Bayesian network cannot establish the hierarchical relationships among hypotheses when constructing its structure. In fact, hypotheses α, β, and γ are in the same level with η, ξ, and τ. Therefore, the two requirements in the Bayesian network cannot be satisfied, resulting in its failure.

Further Readings

How RFID Works

Kevin Bonsor & Wesley Fenlon "How RFID Works" November 5, 2007. HowStuffWorks.com. http://electronics.howstuffworks.com/gadgets/high-tech-gadgets/rfid.htm, accessed on April 9, 2016.

This article presents the types of RFID tags and discusses how these tags can be used in different cases such as in supply chain and in Departments of State and Homeland Security.

User Tracking
http://people.csail.mit.edu/fadel/wivi/index.html
The site provides information about a new technology named Wi-Vi that use Wi-Fi signals to track moving humans through walls and behind closed doors. Wi-Vi relies on capturing the reflections of its own transmitted signals off moving objects behind a wall in order to track them.

Wireless Sensor Networks
http://www.sensor-networks.org/
This is a site for the Wireless Sensor Networks Research Group, which is made up of research and developer teams throughout all the world that are involved in active projects related to the WSN field. The site provides information about a variety of WSN projects.

Context Data Distribution
P. Bellavista, A. Corradi, M. Fanelli, and L. Foschini, A survey of context data distribution for mobile ubiquitous systems. *ACM Comput. Surv.* vol. 44, no. 4(24), pp. 1–45, 2012.
This paper analyzes context data distribution that is especially important for context gathering.

Dempster–Shafer Theory
L. Liu and R. R. Yager, Classic works of the dempster-shafer theory of belief functions: an introduction, *Studies in Fuzziness and Soft Computing*, vol. 219, pp. 1–34, 2008. Springer-Verlag Berlin Heidelberg.
This book brings together a collection of classic research papers on the Dempster–Shafer theory of belief functions.

Context Modelling and Reasoning
C. Bettini, O. Brdiczka, K. Henricksen, J. Indulska, D. Nicklas, A. Ranganathan, and D. Riboni, A Survey of Context Modelling and Reasoning Techniques, *Pervasive and Mobile Computing*, vol. 6, no. 2, pp. 161–180, 2010.
This paper discusses the requirements for context modeling and reasoning.

References

1. G. D. Abowd, A. K. Dey, P. J. Brown, N. Davies, M. Smith, and P. Steggles, Towards a Better Understanding of Context and Context-Awareness, in *Proceedings of the 1st International Symposium on Handheld and Ubiquitous Computing (HUC'99)*, 1999, pp. 304–307.
2. J. W. Kaltz, J. Ziegler, and S. Lohmann, Context-Aware Web Engineering: Modeling and Applications, *Revue d'Intelligence Artificielle*, vol. 19, no. 3, pp. 439–458, 2005.
3. M. Yu, A. Malvankar, and W. Su, An Environment Monitoring System Architecture Based on Sensor Networks, *International Journal of Intelligent Control and Systems*, vol. 10, no. 3, pp. 201–209, 2005.
4. H. Chao, Y. Q. Chen, and W. Ren, A Study of Grouping Effect on Mobile Actuator Sensor Networks for Distributed Feedback Control of Diffusion Process Using Central Voronoi Tessellations, *International Journal of Intelligent Control and Systems*, vol. 11, no. 2, pp. 185–190, 2006.

5. I. F. Akyildiz, W. Su, Y. Sankarasubramaniam, and E. Cayirci, A Survey on Sensor Networks, *IEEE Communications Magazine*, vol. 40, no. 8, pp. 102–105, 2002.

6. A. Boukerche, R. Werner Nelem Pazzi, and R. Borges Araujo, Fault-Tolerant Wireless Sensor Network Routing Protocols for the Supervision of Context-Aware Physical Environments, *Journal of Parallel and Distributed Computing*, vol. 66, no. 4, pp. 586–599, 2006.

7. K. Akkaya and M. Younis, A Survey on Routing Protocols for Wireless Sensor Networks, *Ad Hoc Networks*, vol. 3, no. 3. pp. 325–349, 2005.

8. J. Hightower, and G. Borriello, Location Systems for Ubiquitous Computing, *Computer*, vol. 34, no. 8, pp. 57–66, 2001.

9. G. Roussos and V. Kostakos, RFID in Pervasive Computing: State-of-the-Art and Outlook, *Pervasive and Mobile Computing*, vol. 5, no. 1, pp. 110–131, 2009.

10. J. Ye, L. Coyle, S. Dobson, and P. Nixon, Ontology-Based Models in Pervasive Computing Systems, *The Knowledge Engineering Review*, vol. 22, no. 4, pp. 315–347, 2007.

11. R. Want, A. Hopper, V. Falcão, and J. Gibbons, The active badge location system, *ACM Transactions on Information System*, vol. 10, pp. 91–102, 1992.

12. O. Rashid, P. Coulton, and R. Edwards, Providing Location Based Information/Advertising for Existing Mobile Phone Users, *Personal and Ubiquitous Computing*, vol. 12, no. 1, pp. 3–10, 2008.

13. U. Bandara, M. Hasegawa, M. Inoue, H. Morikawa, and T. Aoyama, Design and Implementation of a Bluetooth Signal Strength Based Location Sensing System, in *Proceedings of the IEEE Radio and Wireless Conference*, 2004, pp. 319–322.

14. A. Harter, A. Hopper, P. Steggles, A. Ward, and P. Webster, The Anatomy of a Context-Aware Application, *Wireless Networks*, vol. 1, pp. 1–16, 2001.

15. N. B. Priyantha, A. Chakraborty, and H. Balakrishnan, The Cricket Location-Support System, in *Proceedings of the 6th Annual International Conference on Mobile Computing and Networking (MobiCom)*, 2000, pp. 32–43.

16. Y. Liu, Y. Zhao, L. Chen, J. Pei, and J. Han, Mining Frequent Trajectory Patterns for Activity Monitoring Using Radio Frequency Tag Arrays, *IEEE Transactions on Parallel & Distributed Systems*, vol. 23, no. 11, pp. 2138–2149, 2012.

17. L. M. Ni and A. P. Patil, LANDMARC: Indoor Location Sensing Using Active RFID, in *Proceedings of the First IEEE International Conference on Pervasive Computing and Communications, 2003. (PerCom 2003)*, pp. 407–415, 2003.

18. C. Qian, H. Ngan, Y. Liu, and L. M. Ni, Cardinality Estimation for Large-Scale RFID Systems, *IEEE Transactions on Parallel Distributed Systems*, vol. 22, no. 9, pp. 1441–1454, 2011.

19. P. Wilson, D. Prashanth, and H. Aghajan, Utilizing RFID Signaling Scheme for Localization of Stationary Objects and Speed Estimation of Mobile Objects, in *IEEE International Conference on RFID*, 2007, pp. 94–99.

20. D. Zhang, J. Zhou, M. Guo, J. Cao, and T. Li, TASA: Tag-Free Activity Sensing Using RFID Tag Arrays, *IEEE Transactions on Parallel Distributed Systems*, vol. 22, no. 4, pp. 558–570, 2011.

21. H. Li, L. Zheng, J. Huang, C. Zhao, and Q. Zhao, Mobile Robot Position Identification with Specially Designed Landmarks, in *Fourth International Conference on Frontier of Computer Science and Technology*, 2009, pp. 285–291.

22. H. Li, S. Ishikawa, Q. Zhao, M. Ebana, H. Yamamoto, and J. Huang, Robot navigation and sound based position identification, in *IEEE International Conference on Systems, Man and Cybernetics*, 2007, pp. 2449–2454.

23. Y. Liu, T. Yamamura, T. Tanaka, and N. Ohnishi, Extraction and Distortion Rectification of Signboards in a Scene Image for Robot Navigation, *Transactions on Electrical and Electronic Engineering Japan C*, vol. 120, no. 7, pp. 1026–1034, 2000.

24. P. Parodi and G. Piccioli, 3D Shape Reconstruction by Using Vanishing Points, *IEEE Transactions on Pattern Analysis and Machine Intelligence*, vol. 18, no. 2. pp. 211–217, 1996.

25. R. S. Weiss, H. Nakatani, and E. M. Riseman, An Error Analysis for Surface Orientation from Vanishing Points, *IEEE Transactions on Pattern Analysis and Machine Intelligence*, vol. 12, no. 12, pp. 1179–1185, 1990.

26. M. Li, M.-Y. Wu, Y. Li, J. Cao, L. Huang, Q. Deng, X. Lin, et al. ShanghaiGrid: An Information Service Grid, *Concurrency and Computation: Practice and Experience*, vol. 18, no. 1, pp. 111–135, 2005.

27. A. P. Dempster, A Generalization of Bayesian Inference, *Journal of the Royal Statistical Society Series B*, vol. 30, no. 2, pp. 205–247, 1968.

28. G. Shafer, *A Mathematical Theory of Evidence*. Princeton University Press, Princeton, New Jersey, 1976.

29. P. Orponen, Dempster's Rule of Combination is NP-Complete, *Artificial Intelligence*, vol. 44, no. 1–2, pp. 245–253, 1990.

30. S. McClean, B. Scotney, and M. Shapcott, Aggregation of Imprecise and Uncertain Information in Databases, *IEEE Transactions on Knowledge and Data Engineering*, vol. 13, no. 6, pp. 902–912, 2001.

31. D. Mercier, B. Quost, and T. Denœux, Refined Modeling of Sensor Reliability in the Belief Function Framework Using Contextual Discounting, *Information Fusion*, vol. 9, no. 2, pp. 246–258, 2008.

32. Y. Ma, D. V. Kalashnikov, and S. Mehrotra, Toward Managing Uncertain Spatial Information for Situational Awareness Applications, *IEEE Transactions on Knowledge and Data Engineering*, vol. 20, no. 10, pp. 1408–1423, 2008.

33. M. Mamei and R. Nagpal, Macro Programming through Bayesian Networks: Distributed Inference and Anomaly Detection, in *Proceedings to the Fifth Annual IEEE International Conference on Pervasive Computing and Communications (PerCom)*, 2006, pp. 87–93.

34. A. Ranganathan, J. Al-Muhtadi, and R. H. H. Campbell, Reasoning about Uncertain Contexts in Pervasive Computing Environments, *IEEE Pervasive Computing*, vol. 3, no. 2, pp. 62–70, 2004.

35. R. R. Yager, On the Dempster-Shafer Framework and New Combination Rules, *Information Sciences (New York)*, vol. 41, no. 2, pp. 93–137, 1987.

36. P. Smets, The Combination of Evidence in the Transferable Belief Model, *IEEE Transactions Pattern Analysis Machine Intelligence*, vol. 12, no. 5, pp. 447–458, 1990.

37. *House_n Project*, Department of Architecture, Massachusetts Institute of Technology. [Online]. Available from: http://architecture.mit.edu/house_n/intro.html, accessed on February 15, 2009.

Chapter 4

Resource Management in Pervasive Computing

Pervasive computing encompasses a broad range of capabilities including communications (e.g., cell phones), mobile computing (e.g., laptop computers), image transferring (e.g., displays), and sensors (e.g., RFID and cameras), where efficient resource management is important for making pervasive systems available anytime and anywhere in the world. More specifically, infrastructures and challenges, from resource allocation to task migration, must be carefully maintained in order to make pervasive systems more efficient.

In this chapter, we introduce ways to allocate resources efficiently and then present how to migrate tasks intelligently in pervasive environments.

4.1 Efficient Resource Allocation in Pervasive Environments

In the current pervasive computing environment, more and more microprocessors with extremely limited resources will be embedded into pervasive devices [1]. We call these devices processing elements (PEs) because each PE only has a limited function. Accordingly, all PEs form a pervasive multiprocessor (PMP) system [2,3]. Different kinds of PEs have different functions, and several kinds of PEs can work together to complete a complex task.

There are many kinds of PEs in a PMP system. The objective of PMP systems is to organize heterogeneous PEs in order to run pervasive applications across the PEs. Therefore, resource allocation is a key issue that must be solved before these PMP systems can be put into practice.

In the following section, we will introduce two kinds of resource allocation policies as part of a detailed introduction to PMP systems.

4.1.1 Introduction to PMP Systems

In the pervasive society, services are filled around the user as easily as oxygen is supplied. This requires computing resources with two opposing attributes: higher performance and lower power consumption. A PMP system consists of many PEs and provides various functions for users by combining certain specified PEs. In the pervasive society, users' needs are multifarious so that the best solution is to offer several basic functions that can collaborate with others to provide various services. This idea brings flexibility dynamics to a PMP system.

In a PMP system, there are three kinds of nodes, as shown in Figure 4.1.

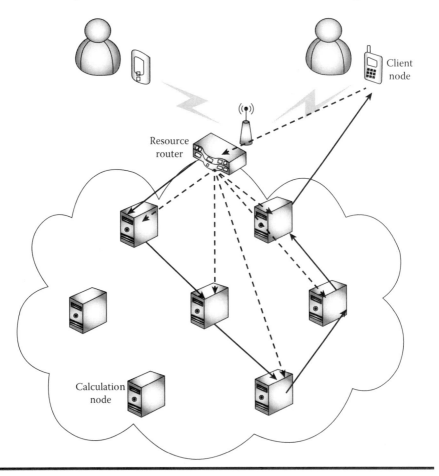

Figure 4.1   A pervasive multiprocessor system.

- *Client node.* This node requests tasks for mobile users through mobile terminals in a wireless network.
- *Resource router (RR).* This is a gateway of the PMP system. There exists only one RR node in one subnet. The RR node receives task requests from the client nodes and determines which tasks should be currently executed in the subnet.
- *Calculation node.* This executes tasks from the client nodes. When a task is executed, several calculation nodes are connected to each other like a chain. To encode a bitmap file into a JPEG file, for example, six calculation nodes are organized as a chain. Application demands determine which calculation nodes will be combined.

The core components of PMP systems are illustrated in Figure 4.2, and they interact as follows. First, in user space, a pervasive application sends a task to *task management* via a ubiquitous service broker (USB). Then, task management asks the *task analysis and decomposition* component to decompose the task into small subtasks. Each subtask can be completed by a corresponding PE. After decomposing, the task layer sends the subtasks to the service layer. Components in the service layer will require PEs from the resource layer and will schedule the PEs for subtasks. The resource layer is responsible for managing PEs, including monitoring PE status.

The PMP system works in the following way. A client uses a mobile phone to take a photo, which is in raw form and takes up a large space. Then the pervasive application on the mobile phone connects to the RR node via a fixed UDP port. After connecting, the RR spawns a new client thread to communicate with the client (here, this refers to the mobile phone) for the current task. Then the mobile phone sends the task (here, this refers to JPEG encoding) and decomposed subtasks to the new client thread. The new client thread finds all the necessary PEs for these subtasks. If it cannot find them all, it rejects the task. If it finds them all, it sends the subtasks to the PEs, and any of these PEs cannot be allocated again for other tasks until all of these PEs are freed by the system. Finally, the PE in the last step sends the result to the mobile phone.

4.1.2 Pipeline-Based Resource Allocation

The primitive resource allocation scheme in PMP systems does not release all reserved PEs until the whole task is finished, which causes a huge waste of computational resources. To mitigate such a limitation, we designed two pipeline-based algorithms: a randomly allocating algorithm (RAA) and an RAA with cache (simplified as RAAC), in which the RR only allocates a necessary PE during the current phase. First, we review the primitive resource allocation algorithm (i.e., the current algorithm, or CA) as follows.

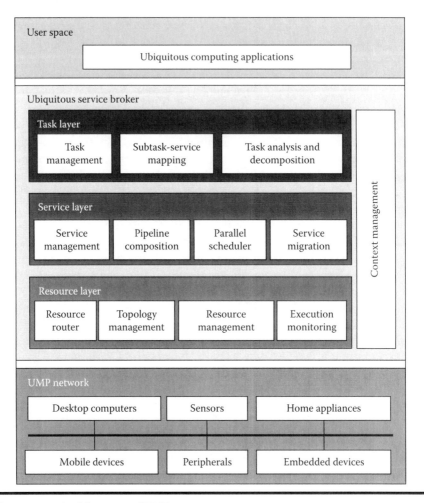

Figure 4.2 Architecture of PMP systems.

In the CA, when a task arises, the RR will reserve all the PEs needed for the task. During the processing, all allocated PEs cannot be used by other tasks even when they are free. In the CA, the total delay can be calculated as follows:

$$d = \frac{m \cdot 0t + m \cdot 1t + m \cdot 2t + \ldots + m \cdot \left(\left\lfloor \dfrac{n}{m} \right\rfloor - 1\right)t + \left(n - \left\lfloor \dfrac{n}{m} \right\rfloor m\right) \cdot \left\lfloor \dfrac{n}{m} \right\rfloor t}{n} \quad (4.1)$$

where m is the number of tasks that RR can handle at one time, and t is the time to handle the m tasks. So, the ith m tasks wait $(i-1)$ time units.

Task execution efficiency can be formulated as follows:

$$f = \frac{\sum_{i=1}^{n} e_i}{\sum_{i=1}^{n} e_i + \sum_{i=1}^{n-1} c_{i,i+1}}$$

(4.2)

where f refers to task execution efficiency, e_i ($1 \le i \le n$) is the execution time in ith PE, and $c_{j,j+1}$ is the communication time between jth PE and $(j+1)$th PE, $1 \le j \le n$.

To improve the task execution efficiency in CA, we design an RAA. The basic idea of an RAA is that after the PE is finished executing the process, the PE will ask the RR for the next phase. Task execution efficiency can be formulated as follows:

$$f = \frac{\sum_{i=1}^{n} e_i}{\sum_{i=1}^{n} e_i + cr_1 + \sum_{i=2}^{n} \left(2 * cr_i + c_{i-1,i}\right)}$$

(4.3)

where e_i ($1 \le i \le n$) is the execution time unit, cr_i ($1 \le i \le n$) is the communication time between ith PE and RR, and $c_{i-1,j}$ ($2 \le i \le n$) is the communication time between $(i-1)$th PE and ith PE.

For mobile users, the battery lifetime is a very important factor in the system design. To reduce the energy consumption on the user side, we adjust the first PE and the last PE to provide frequent access for the user to search for the last PE. Thus, all the optimization process has an effect on the middle PEs in the whole process chain.

In the RAA algorithm, the usage rate of a PE is quite high, but the load in the RR is heavy. To reduce the load in the RR, we introduce a cache technology and design, the RAAC algorithm, where we assign a cache for every PE to memorize the next stage's PE. When a PE finishes its subtask, it will search the next phase of PE in its cache. If all PEs in the cache are at a busy status, it will ask RR to assign one free PE as the next phase PE. In this algorithm, task execution efficiency will become

$$f = \frac{\sum_{i=1}^{n} e_i}{\sum_{i=1}^{n} e_i + \sum_{i=2}^{n} \left(3c_{i-1,i}\right)}$$

(4.4)

where e_i ($1 \le i \le n$) is the execution time unit, cr_i ($1 \le i \le n$) is the communication time, and $c_{i-1,j}$ ($2 \le i \le n$) is the communication time between $(i-1)$th PE and ith PE.

In resource allocation for PMP systems, another important concern is how to allocate PEs when there are no available qualified PEs because each PE is deployed for a specified kind of task (e.g., JPEG encoding). In addition, although all users request the same tasks, there are not enough available PEs, and some of the requests are rejected. To solve this problem, we designed the following two kinds of PEs: *available PEs* and *empty PEs*. An available PE executes a specified code only for a corresponding subtask. In order to complete a task, a group of available PEs is combined for all the subtasks in a task. On the other hand, an empty PE itself does not carry any code; however, it can load any kind of codes from a repository PE or a repository server so that it can perform any corresponding subtask.

We utilize empty PEs when we cannot find any available PE, when available PEs are all busy, or when they cannot deal with a requested task. In summary, PEs can be allocated for handling a diversity of tasks as follows.

1. Discover a set of available PEs. (This does not mean they all are allocated at one time, this is just for testing the requirement.) If all are found, that is fine; otherwise, if all cannot be found but there are empty PEs, proceed to Step 2. Otherwise, reject the task.
2. Ask the system for necessary codes, if the codes are loaded to empty PEs, go to Step 3.
3. Ask the client for necessary codes (this is handled by the pervasive application and is unseen by any user), if the codes are loaded to empty PEs, go to Step 4.
4. Check if the task is a migrating task; if so, borrow the code from the last location; if not, reject the task.

Users are unaware of the above steps. So, the users experiment with better performance without being rejected by the system.

We built a simulation system to evaluate the aforementioned three algorithms with a Poisson distribution [4] based task arrival ratio. The Poisson distribution is applicable to various phenomena of a discrete nature whenever the probability of a phenomenon happening is constant in time or space. The number of tasks is from 2500 to 5500. We run the simulation of the number of tasks from 2500 to 5500 every 150 intervals. We set the number of PEs as 2400. Each task needs six PEs to process the six JEPG encoding steps, respectively. Therefore, the total number of PE chains is 400. We also set the network delay as 100.

Figure 4.3 shows the load balance in RAA and RAAC schemes. The load balance of CA is obviously smaller than that of the RAA and the RAAC because once the RR assigns the PE to execute the task, it will never communicate with the PEs. From this image, we can see that, by using the cache technology, the load balance in the RAAC has a better result than that of the RAA. The reason is that almost every time the PE should ask the RR for the next phase PE.

Task execution efficiency greatly depends on waiting time. As shown in Figure 4.4, CA shows the worst result because it has to wait for the execution to

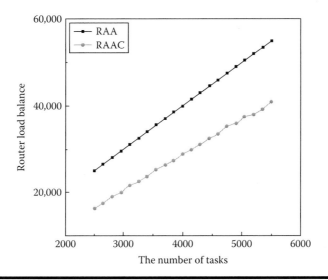

Figure 4.3   Load balance in RAA and RAAC.

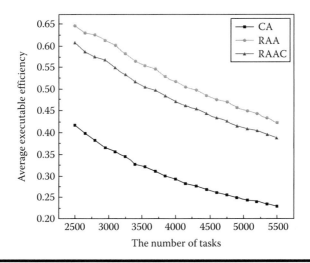

Figure 4.4   Tasks' execution efficiency in CA, RAA, and RAAC.

start even though there are free PEs in the process chain. The RAA outperforms the RAAC; however, it has a serious load balance problem.

Delay (waiting time) is an important factor in a real system. Even if the total execution time is good, if the delay is very large, the system still is not efficient for users. From Figure 4.5, we can see that the average of CA delay is extremely large. That is because the execution procedure is almost sequential. The reason why the RAA is

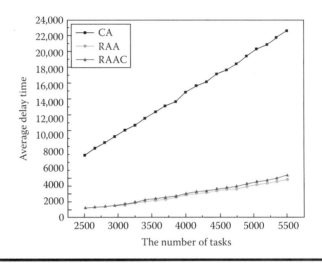

Figure 4.5 **The delay (waiting time) in CA, RAA, and RAAC.**

slightly better than the RAAC is the RAA can fully randomly use the next phase of PE. Therefore, the waste from failed communication time is omitted.

Through these experiments, we have successfully proven the value of our proposed algorithms. Considering these key factors (task execution efficiency, load balance of the RR, the delay of task execution), the RAA and RAAC have both merit and demerit. There is always a trade-off between these factors. Taking into account the balance point of all factors, we found the RAAC is much more suitable in the real environment for allocating resources (PEs) when the system has many users and many tasks to process. If the users and tasks are not that large, the RAA could perform well, too. One other thing we have realized through these experiments is that we can set the allocating policy flexibility based upon the user's requests.

4.1.3 Probabilistic Approach-Based Resource Allocation

To further improve the efficiency of resource allocation in the aforementioned PMP systems, a probabilistic approach can be adopted, which makes the PMP system able to look for an item that needs to be processed earlier and more efficiently. Moreover, it can omit a useless retrieval that causes a waste of energy.

4.1.3.1 Basic Algorithm

First, in the following steps, we introduce the basic algorithm through which the RR searches idle PEs.

- **Step 1:** The RR searches an idle PE randomly and sends a PE a message to check its state.

- **Step 2:** The PE sends the RR an answer of idle/busy.
- **Step 3:** After the RR gets the answer of idle, it finishes the search. If the answer is busy, it returns to Step 1.

Specifically, the RR randomly searches for a PE in an idle state by sending a message asking a PE whether it is in an idle state or a busy state. The PE replies to the RR with its current state when it receives the message. When the PE responds that it is in an idle state, the RR sends the PE a message requesting execution of the task. Otherwise, the RR looks for another available PE. It keeps retrieving idle PEs until all the tasks are executed.

We can consider two possible scenarios for the RR to search for an idle PE: the best case and the worst case. The best case means the RR successfully finds an idle PE in the first search (i.e., the first retrieved PE is in an idle state, so the RR no longer looks for another PE). By contrast, the worst case means the RR finds an idle PE only after searching every PE on the subnet (i.e., it retrieves the idle PE at last after finding out that all of the other PEs are busy).

In the basic algorithm, the RR looks for an available PE at random. Therefore, the best case cannot always be expected, and the worst case cannot be avoided. One reason for the worst case is that PEs are often in a busy state. A long delay can occur if the RR keeps retrieving busy PEs instead of idle ones.

We consider that if the RR omits retrieving such PEs, it can find an idle PE quickly and efficiently. The basic idea can be described as follows: (1) first forming a group of PEs that are often in a busy state and (2) the RR does not retrieve any PE from that group. A concrete method is needed for finding such busy PEs and is presented in the following subsection.

4.1.3.2 Optimized Algorithm

In the optimized algorithm, the RR first searches for an idle PE randomly and checks each PE's state. After the RR repeats this procedure many times, it forms a group of busy PEs based on a history of each PE's state when the RR checked. In the next search, the RR tries to find an idle PE excluded from this group. If the RR cannot find any idle PE, it searches for an idle PE from inside and outside of the group. This algorithm is based on an assumption that there exists at least one idle PE out of all PEs. Details of the algorithm are described in the following steps:

- **Step 1:** The RR secures a storage region in each PE to save its degree of idle state, which starts from 0.
- **Step 2:** The RR searches for an idle PE randomly and sends that PE a message to check its state.
- **Step 3:** The PE sends the RR an answer of idle/busy.
- **Step 4:** If the RR gets the answer of idle from the PE, it finishes the search and the PE's degree of busy state is decremented by 1. If the answer is busy, the PE's degree is incremented by 1, and the RR searches the next PE.

- **Step 5:** If repeating from Step 2 to Step 4 s times, the RR looks into each PE's storage. If a PE's degree of busy state is more than a threshold, the PE is regarded as a *PE with a high probability of busy.*
- **Step 6:** After looking into all the PEs' storage, the RR forms a group of PEs with a high probability of busy.
- **Step 7:** The RR searches an idle PE again, except for PEs belonging to the group in Step 6, t times. If the RR cannot find any idle PE, it begins to search for an idle PE including inside of the group.
- **Step 8:** Steps 5 through 7 are repeated until the entire search ends.

In the optimized algorithm, we can also consider the best case and the worst case for the RR spending time searching for an idle PE. The best case can occur when the RR finds an idle PE during the first search, just like the best case of the basic algorithm. On the other hand, the worst case is if the RR first searches an idle PE from outside of the group but all PEs are in the busy state. Hence, the RR starts to look for an idle PE from the group and then finally finds one after the all other PEs in the group are queried.

4.1.3.3 Probability Analysis

We evaluate the optimized algorithm using probabilities of finding an idle PE in the best case and the worst case, respectively. Both depend on probabilities of a PE in an idle state and a busy state, respectively.

The queuing theory can be used to stochastically analyze congestion phenomena of queues and allows us to measure several performances such as the probability of encountering the system in certain states, such as idle and busy. We use the queuing theory [5] to obtain probabilities of a PE in an idle state and a busy state. Here, we consider that the queuing model of PEs can be described as M/M/1(1) by Kendall's notation [6]. That means each PE receives a message from the RR randomly and deals with only one task at a time. Hence, when a PE processes a task and a message comes from the RR at same time, the message is not accepted by the PE due to its busy state.

In the best case, the probability of finding an idle PE is

$$P = P_0 \tag{4.5}$$

In the worst case, the probability of finding an idle PE is

$$P = P_1^{(n-1)} \tag{4.6}$$

On the other hand, when the optimized algorithm is used, probabilities in the best case and the worst case are Equations 4.7 and 4.8, respectively.

$$P = P_0 \tag{4.7}$$

$$P = P_1^{(n-m)} \cdot P_1'^{(m-1)} \tag{4.8}$$

The best case is the same as the one using the basic algorithm because the RR finds an idle PE at the first search in both algorithms. Therefore, we consider only the worst cases when comparing the optimized algorithm with the basic algorithm.

Based on this assumption, we obtain the probabilities of the worst case in the basic algorithm and in the optimized algorithm using Equations 4.6 and 4.8, respectively. The greater the total number of PEs, the larger the difference in the probabilities becomes between the basic algorithm and the optimized algorithm. Therefore, we can conclude that the optimized algorithm has less risk of encountering the worst case than the basic algorithm because the total number of PEs is larger. Finally, the number of PEs in the group has less of an impact on results when compared to the other two parameters.

As a result, when using the optimized algorithm the worst case is encountered less often than when using the basic algorithm if the following conditions are met: (1) the threshold for making a busy group by checking each PE ranges from 0.6 to 0.8 and (2) the total number of PEs in a subnet is relatively large. Considering these two conditions in implementation, effective performance using the optimized algorithm can be expected.

4.2 Transparent Task Migration

In pervasive applications, tasks often need to be transparently migrated among extremely heterogeneous platforms with the movement of pervasive users. As a result, context-sensitive task migration is an important enabling technology to achieve the attractive human-centric goal of pervasive computing. To achieve online task migrations, we have to redirect resources, recover application states, and adapt multiple modalities.

4.2.1 Introduction to Task Migration

Task migration technologies can be categorized as *desktop level migration, application level migration*, and *process level migration*, which each support different levels of remote applications. In desktop level migration (e.g., XWindow [7], VNC [8], and Windows Remote Desktop [9]), the whole remote windows/desktop visualization is brought locally. This migration model lacks control of applications and is not aware of local computing contexts (e.g., local file systems, resources, devices, and services).

Application level migration can be subdivided into two categories of methods. One is the naïve method that packages original installation files and then transfers these files to the target workstation. Many enterprise management tools already employ this method to customize and deploy applications [10]. Another method is built on the C/S pattern, utilizing the original workstation as a server and the migration-target platform as a client. The application level user interface (UI), rather than the whole desktop bitmap, is migrated to and re-rendered to the target host. By comparison, the second method is more flexible and plastic and has little cost in pervasive computing contexts.

Process level migration focuses on the kernel (i.e., migrating some active execution images from a source computing context to the target). Process migration can be tailored to migrate only some subsets of computing contexts, and thus exhibits the most flexibility. Its primary complexity lies in the kernel level status maintenance and platform heterogeneity [11].

In a pervasive system, the task migration method should efficiently handle the following characteristics:

- *Heterogeneous software and hardware platforms.* Not only video players and operating systems but also devices are completely different before and after the migration.
- *Delay-sensitive migration process.* Too much delay is always undesirable to users.
- *Lightweight task migration.* There should be as little migrated data as possible because of the limited bandwidth of wireless networks and the prompt response requirement.
- *Multimodality representation.* After the migration, a lower-resolution movie is preferred, adapting to a small screen as well as to the lower bandwidth of wireless networks.

In the following section, we introduce a shadowlike task migration model (called xMozart) that can migrate tasks in a shadowlike way and that can automatically adjust execution and display behaviors based on run-time context. This migration scheme is based on the Mozart Programming System [12], which supports multiplatform (UNIX and Windows) programming for intelligent and distributed applications.

4.2.2 Task Migration Model

The xMozart model modifies and extends the Mozart platform, with the following key technologies.

4.2.2.1 HTML/DHTML Viewstate

An HTML/DHTML viewstate [13] is a browser-based approach for expanding the states of the Web controls in Web form view. All control states and values are serialized into a base64 string, which is then put into the viewstate field via a Web page. The viewstate file will be sent back to the Web server as the page posts back. The Web server will compose a hierarchical tree of all the controls in the Web page, then de-serialize the viewstate field and apply the viewstate to the Web controls. The Web server reconstructs the Web application and recovers the states of the Web controls. This process realizes the Web application migration from the client side to the server side.

4.2.2.2 OZ Source File Reorganization

OZ [14] is the official language used in Mozart. It is a multiparadigm programming language that supports declarative programming, object-oriented programming, constraint programming, and so on. GUI programming in Mozart is supported through a QTK toolkit [15], an extension of the TK [16] module that offers an object-oriented binding to TCK/TK. QTK takes advantage of OA records to introduce a partly declarative approach to programming user interfaces. Based on the Web page viewstate approach, we tuned the OZ program paradigm so that it can succeed our requirements for application migration.

As illustrated in Figure 4.6, besides the normal OZ source records, we added two other sections (a state description section and a resource description section) in the migrating process to service state persistency and resource transformation. After migration, the migrated file will be sent to our framework to be further scrutinized, during which time the state/resource section will be de-serialized and mapped into normal OZ records to form the initialization section. Meanwhile, the two persisted sections will be removed from the resulted to-be-executed OZ source file.

The first execution after migration will begin with an initialization section to recover control states and rebinding resources, after which the execution continues at the normal OZ source section, and the application resumes on a new workstation from the paused point on the original machine.

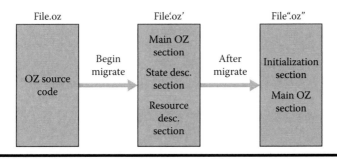

Figure 4.6   OZ source file reorganization.

4.2.2.3 Compiler Support

We tailored the OZ compiler to make it accept our newly added language semantics.

4.2.2.3.1 Section-Based OZ Programming

The classic OZ programming paradigm is a flat programming model. All OZ source files contain only OZ records (widgets, etc.). Our new programming model introduces three other sections: a state description section, a resource description section, and an initialization section (with normal OZ records in the main OZ section). Now, the OZ compiler may accept four program sections.

As illustrated in Figure 4.7, when an OZ application is ready to migrate, the framework will raise an interrupt. First, the migration module will extract all the variables indicated as resource handles, perform the LRI to URI transformation, and then it will serialize the resource handles to the resource section in the OZ source file. Second, all the variables indicated as state variables will be extracted and serialized to the state section of the OZ source. At this time, the original execution halts and migration begins by sending the amended OZ source file to the destination platform, where the modified OZ compiler will go through the migrated OZ source file, de-serialize the state section and the resource section, and then will auto-generate a mapped OZ *setter* clause to set the variable state to vary accordingly and will rebind the transformed resource handle to the resource variable, respectively.

In the main OZ section, a compiler will disable all the initialization/assignment operations that may overwrite our variables initialization in the initialization section. Now, the execution may resume by processing the compiler customized OZ source file sequentially.

4.2.2.3.2 New Semantic for Resource Handles

To indicate resource variables, we introduce a new meta-word `Declare ImageHandler {[ResourceAttribute],...}`. We modify the OZ compiler to accept this keyword when parsing OZ source files. With this form of resource declaration, the migration module of the framework will be notified to transform and initiate serialization when a migration is triggered.

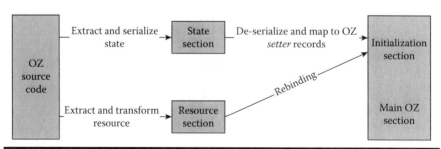

Figure 4.7 OZ sections.

4.2.2.3.3 New Semantic for State Variables

Similarly, to indicate the state variables, we introduce a new meta-word `Declare TruckPosition {[StateAttribute], x,y}`. We modify the OZ compiler to accept this keyword when parsing OZ source files. With this form of state declaration, the compiler will generate some shadow function automatically, which may be called the migration module when a migration is triggered.

4.2.2.3.4 New Semantics for Modality Interchange

Another semantic we introduce in our framework is the meta-word `Declare InputModule {[Interchangeable]}`, which may be used to indicate that some particular I/O module is subject to supplementation or substitution.

4.2.2.3.5 Initialization Section

This is a generated section for rebinding resources that is a recovery state after migration. This section is made up of purely normal OZ records, which are auto-generated by the OZ compiler when de-serializing resource and state information from the *hidden* part of the OZ source files. Typically, this section is composed of OZ record *setters*: `Set TruckPosition={100,200}`. This section suppresses all the variable definitions, initialization, and assignment in the original OZ program, which is, of course, supported by the compiler.

4.2.2.4 Application State Persistence and Recovery

4.2.2.4.1 Application States and Variable States

The migration process of a particular application covers application API migration, application state migration, original resources redirection, and application state recovery. We define application states as the collection of all state variables. And state variables are those whose states are vital to the process of execution resumption after they are migrated to a new station. State variables fall into two categories: local state variables and global state variables. To migrate application states, we are committed to recover both of these state variable categories.

Global variables have the advantage that every code snippet can access them, thus making our work with extract states easier. We employ a compiler-generated procedure to fetch all global states and then serialize the states to the variable states section. Things become complicated when it comes to the local state variables because we cannot provide a similar procedure for extracting all local state variables. (Technically, we cannot access a local variable from another procedure without passing an argument.) Therefore, we introduce the shadow function technique to cope with the local state variables situation.

4.2.2.4.2 State Persistence and Shadow Functions

Shadow functions are those compiler-generated procedures that help with the state serialization process. If the interruption for migration occurs before those local variable states are written into the state section, we only have the variable states of the last iteration. After migration, execution will resume from the very beginning of the interrupted iteration. Though the code between the very beginning and the interruption point may have to be executed one more time, generally, this approach is acceptable.

```
For (int i=0; i<10; i++){
  ...//normal execution
  //Shadow function goes here to persist the i variable.
  }
```

Regarding the global state variable states' persistence, we provided a compiler-generated procedure as an interrupt handler to handle the uniform persistence to the state sections. The procedure will be called upon every time a migration request is triggered.

4.2.2.4.3 State Recovery and OZ Setters

We need to recover those state variables before continuing application execution on the new workstation. We make the initialization section a uniform place for recovery of all the global state variables. We generate the 1–1 mappings between the global state variables/values from the persisted section and the target OZ program records. That is, for each (variable, value) pair in the state section, a corresponding OZ *setter* record will be generated by the compiler (e.g., TruckPositionX=50). This is a valid OZ source record and should require no extra compiler effort to integrate it into the OZ main source program. All the generated OZ setters are in essence the state recovery methods for recovering the global state variables. They reside in the initialization section and will be executed sequentially the first time the execution runs on the new station.

Similar to the persistence process, local state variable recovery is more challenging. We also employ the code injection technique to complete the local variable recovery.

```
int i=5; //code injection here to recover variable i's state
  For(;;i<10;i++){// the original I assignment will be suppressed
  ...//normal execution
  }
```

During the state recovery of variables, we also suppress all the initialization/assignment OZ clauses to let the state recovery records take effect.

4.2.3 Resource Redirection

4.2.3.1 Resource Handle Transformation Schema

We propose the new semantics for dealing with resources. We added the
`[ResourceAttribute]` tag to denote that the marked resource handle should
be properly transformed during the migration process. One principle for application migration is that we never migrate resources. Resources such as image handles
used by the application on the original station should be also accessible from the
new workstation where the migrated application resides. Therefore, we must provide a way to identify the referenced resources.

We utilize a natural transformation strategy to fulfill this goal by dividing all
the resources used in the original application into two categories: one is the globally accessible resource and the other is the locally accessible resource. The globally
accessible resources are those referenced in the URI form. For example, an image
resource in URI form may look like:

```
//HostServer/Resources /truck.gif
```

where HostServer is the server accessible from the network, with either the
server's network name or the form of IP address. We insert this kind of resource
intact and assume that, if the URI form resources are accessible in the original
station, they can be accessed from the migrating station and also from other
stations that have network connectivity accepting the migration from the original station.

The other resources that we primarily deal with are those in LRI form, which
are those resources on the workstation that are executing the OZ application. An
LRI resource may be in an absolute path form such as:

```
C://Resources/truck.gif  (/Resources/truck.gif on Linux) or
  Resources/car.gif
```

Regarding the LRI form of resource, we add some platform support to realize
the LRI to URI transformations. That is, we make those LRI resources accessible
from the network.

4.2.3.2 Resource Transformation Schema

We define our resource transformation schema as:

```
If oldResourceHandle is of URI form transformedResourceHandle=
  oldResourceHandle;
else {
If oldResourceHandle is of relative LRI form
  oldResourceHandle
```

```
=absolutePathPrefix+oldResourceHandle;
       transformedResourceHandle
="//"+originalHostNameOrIpAddress+oldResourceHandle;
}
```

To make the resource reachable in the transformed form from the migrated station, there must be some mechanism to map the transformed resource handle's link to the local resource path. Our framework provides a module for requesting access to the LRI resources. The module intercepts the incoming resource resolution request from the new workstation, maps it with the local LRI, and then returns the resource. In this sense, the module acts as a proxy between the original station and new station that is handling the resource resolutions (Figure 4.8).

4.2.4 xMozart: A Novel Platform for Intelligent Task Migration

Besides compiler support, the novel xMozart platform is designed to support two aspects of intelligent task migration: the migrate modules that trigger the migration and the modality adapters that support the interchangeable user interface.

4.2.4.1 Migration Management Modules

Migration management modules are vital in our framework for initiating an application migration process. As Figure 4.9 illustrates, an OZ execution can be interrupted to trigger the migration process. The migration trigger handler will halt the current program execution and then the state serialization module will guide the program execution to a compiler-generated routine (the shadow functions) that will be in charge of collecting the resource variables and state variables and serializing them into the resource section and state section, respectively. Then, our framework will prepare the endpoint network for migration to take place.

After the state-enabled OZ source file successfully migrates to the target workstation, the framework will prepare for program resumption by using the deserialization module to extract the resource information and variable states from the resource section and the state section. This will be used as input for the state recovery module, which will process the resource transformation and state variable mappings with OZ grammars. Once all this is accomplished, the halted program is ready to continue execution.

4.2.4.2 Multiple Modalities

Multiple modalities support is the feature we invented to embrace pervasive context computing. This deals with environment-sensitive and device-sensitive contexts. We empower our framework for context-sensitive modality by providing two key modules: the device discovery and evaluation module (DDEM) and the I/O adapter management module (IOAMM). We separate the program I/O module from the

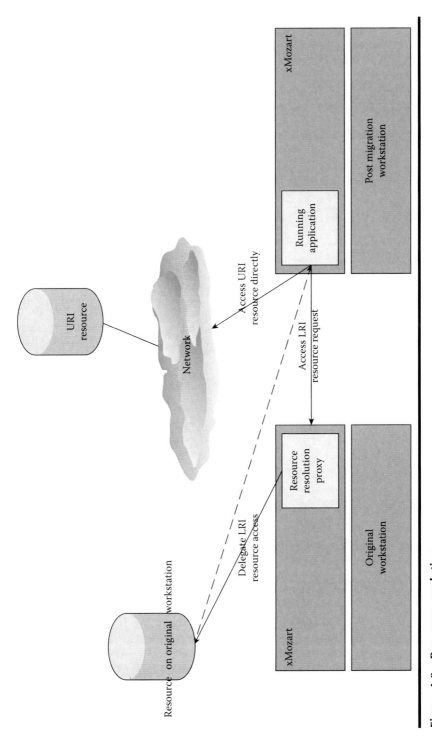

Figure 4.8 Resource resolution.

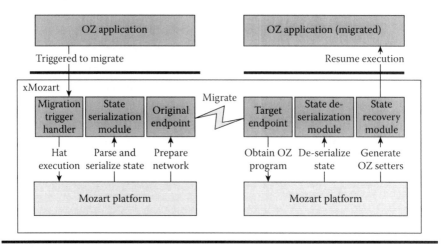

Figure 4.9   Migration management modules.

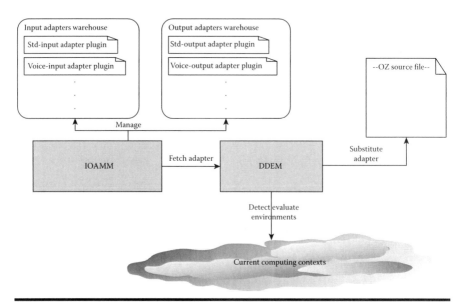

Figure 4.10   Multimodality support.

main OZ program routines so that the I/O module can be substituted as necessary or as the DDEM module deems suitable.

In Figure 4.10, IOAMM manages all the platform-supported I/O interface adaptors. I/O interface adaptors can be added or extended in the form of plug-ins. For instance, our platform supplies input to the main OZ program by providing the standard input module, which obtains input data from the keyboard and feeds the main program. Under this framework, we can substitute the standard input module by providing a voice input adapter, which may be more appropriate in some pervasive computing contexts and may result in a more improved user experience.

DEM is the core module making that decides whether some particular I/O module is available and which is the most proper module to use. DDEM will routinely enumerate all the potential I/O capabilities of the computing environment and make a recommendation. DDEM will then notify the user to shift to another I/O device. With permission granted, DDEM may remove the original I/O adapters and replace them with the proper ones for a specific context. The substitution is subject to modifying the OZ source files, which makes the whole program interface shiftable.

4.2.4.3 Multimodal Programming in xMozart

Applications migrated in pervasive contexts are detached from user interfaces. I/O interfaces used in the original workstation may be not appropriate or even unavailable in the new contexts. For example, the keyboard input method in a personal computer may be considered inconvenient if the application migrated to a driving car, where a voice command input would be preferable.

The xMozart platform provides application of multimodalities from two respects. On the OZ source program level, the I/O modules of the source are extracted so they are stand-alone and then highly detached from other parts of the program, with the only interaction through those code snippets known as the standard adapters. This support is fundamental to our goal of achieving the context-aware I/O modules' intelligent substitution. Management of standard adapters is handled by the platform support (via IOAMM). The framework (DDEM) will evaluate the contexts and supplement or substitute a more proper interface if it considers this necessary.

The DDEM framework is designed to be in a conservative mode. That is, I/O interfaces are not supposed to be interchangeable unless an explicit trigger of a potential modal interchange is applied. We use this approach for the sake of safety issues and to avoid the risk of any arbitrary modal substitution that may turn against the users' wishes.

In order to make an application shiftable, the application's main program must be allowed to strictly detach from all the I/O modules, with only the adapter as a pipe to input and output data. Meanwhile, the I/O module must implement the general interface and be denoted with the tag [Interchangeable].

4.2.5 Implementation and Illustrations

We carry out two prototypes based on our new platform. One is to justify the application migration effect and another is to exhibit the application's multimodalities.

4.2.5.1 Prototype for Application Migration

4.2.5.1.1 State Recovery

We employ a classic OZ program from Mozart platform release, i.e., the TruckRace program [11]. Table 4.1 illustrates item types and denotations used in the TruckRace program. We add some new ingredients to the original program—a piece of music (TruckRace.mp3)—that will play as the trucks race forward. We make the truck picture (Truck.gif) and the music file the resource, denoted by the [ResourceAttribute] tag, and we make the race time (the music playing time) and the truck positions the state variables, denoted by the [StateAttribute] tag.

We feed our modified OZ program to the extended Mozart platform and trigger this application to migrate at an arbitrary time after the race starts. The proxy of our xMozart resolves the resource successfully, and the application migrates from the original desktop workstation to the new laptop workstation, as in Figure 4.11. With the resource fetched and program state recovered, the race continues on the new station.

4.2.5.1.2 Migration Latencies

The latency between applications halts on the original station and successfully migrates to the target station; continuing execution is an important factor that we

Table 4.1   Item Types and Denotations in TruckRace

Items	Type	Denotation
TruckRace.mp3	Resource	[ResourceAttribute]
Truck.gif	Resource	[ResourceAttribute]
Race time (music time)	State	[StateAttribute]
Truck positions	State	[StateAttribute]

Figure 4.11   State recovery.

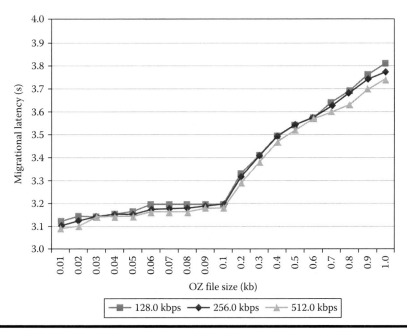

Figure 4.12   Migration latencies.

must take into consideration. This factor is correlated with the size of the modified OZ source file, TruckRace'.oz', network bandwidth to transfer the file to the target location and the platform/compiler work within the migration.

As Figure 4.12 illustrates, we conducted the experiment in network environments with a bandwidth of 128.0, 256.0, and 512.0 kbps. And we fed the OZ source file of the size in the range [0.01 kb, 1.0 kb] into which we thought common applications should fall. The result shows that network conditions today should not be a bottleneck for migration latencies and that the OZ file size (though minor) plays the main role in causing the latencies.

4.2.5.2 Prototype for Multimodalities

We built our second prototype to exhibit the ability of our platform to support multimodalities of applications. We equipped the migration-target workstation with the vocal device, and we pre-installed two well-written plug-ins. One is the standard input adapter through which the input from the keyboard/mouse will feed the application, and another is the vocal command adapter, which can be used to substitute for the standard input adapter after the migration to input command to the application. Those two adapters are successfully discovered and managed by IOAMM.

We triggered the migration and noticed that DDEM discovered the vocal device successfully and collaborated with IOAMM to substitute the original standard input adapter with the vocal adapter. We gave the voice command *start* in the microphone, and the truck started to race after recognizing the command. The command *stop* stopped the race successfully.

Further Readings

Computing Resource Management Systems and Methods
United States Patent Application 20100169893
http://www.freepatentsonline.com/y2010/0169893.html

Predictive Resource Management for Wearable Computing
D. Narayanan and M. Satyanarayanan. Predictive Resource Management for Wearable Computing, in *Proceedings of the 1st international conference on Mobile systems, applications and services* (MobiSys'03). ACM, New York, 2003, pp. 113–128.

XML Systems for Intelligent Management of Pervasive Computing Resources
D. Alexopoulos, G. Kormentzas, and J. Soldatos. XML Systems for Intelligent Management of Pervasive Computing Resources, in *Artificial Intelligence Applications and Innovations: 3rd IFIP Conference on Artificial Intelligence Applications and Innovations (AIAI) 2006*, I. Maglogiannis, K. Karpouzis, and M. Bramer, Eds. Boston, MA: Springer, 2006, pp. 245–253.

Lease-Based Resource Management in Smart Spaces
M. Jurmu, M. Perttunen, and J. Riekki. Lease-Based Resource Management in Smart Spaces, *Fifth Annual IEEE International Conference on Pervasive Computing and Communications Workshops* (PerComW'07), 2007, pp. 622–626.

A Context-Aware Learning, Prediction, and Mediation Framework for Resource Management in Smart Pervasive Environments
N. Roy, "A context-aware learning, prediction and mediation framework for resource management in smart pervasive environments," Ph.D Thesis. Computer Science & Engineering, The University of Texas at Arlington, 2008.

Automatic Resource and Service Management for Ubiquitous Computing Environments
S. Maffioletti, M. S. Kouadri and B. Hirsbrunner. Automatic resource and service management for ubiquitous computing environments, *Pervasive Computing and Communications Workshops, 2004. Proceedings of the Second IEEE Annual Conference on*, 2004, pp. 219–223.

References

1. M. Satyanarayanan, Pervasive Computing: Vision and Challenges, *IEEE Personal Communications*, vol. 8, no. 4 pp. 10–17, 2001.
2. M. Kubo, B. Ye, A. Shinozaki, T. Nakatomi, M. Guo, UMP-Percomp: A Ubiquitous Multiprocessor Network-Based Pipeline Processing Framework for Pervasive Computing Environments, in *Proceedings of the IEEE AINA'07.*

3. A. Shinozaki, M. Shima, M. Guo, and J. Kubo. A High Performance Simulator System for a Multiprocessor System Based on Multi-Way Cluster, in *Proceedings of 2006 Asia-Pacific Conference Computer System Architecture Systems*, Shanghai, China, 2006, pp. 231–243.
4. L. Kleinrock, *Queueing Systems Volume 2: Computer Applications*, Wiley, 1979.
5. Queuing Theory Basics. Available from: http://www.eventhelix.com/RealtimeMantra/ CongestionControl/queueingtheory.htm August 21, 2010.
6. Kendall's Notation. Available from: https://en.wikipedia.org/wiki/Kendall%27s_ notation last modified on September 23, 2015.
7. X.Org project: http://www.x.org/, accessed on October 21, 2010.
8. T. Richardson, Q. Stafford-Fraser, K. R. Wood, and A. Hopper. Virtual Network Computing. *IEEE Internet Computing,* vol. 2, no. 1, pp. 33–38, 1998.
9. Remote Desktop Protocol. Available from: http://msdn.microsoft.com/en-us/library/ aa383015(VS.85).aspx, accessed on December 20, 2015.
10. Microsoft System Center Configuration Manager. Available from: http://www. microsoft.com/systemcenter/configurationmanager/en/us/default.aspx, accessed on August 21, 2010.
11. J. M. Smith. *A Survey of Process Migration Mechanisms.* Technical Report CUCS-324-88. Computer Science Department, Columbia University, New York, NY.
12. The Mozart Programming System. Available from: http://www.mozart-oz.org/, accessed on August 21, 2010.
13. S. Mitchell. Understanding ASP.NET Viewstate. Available from: http://msdn. microsoft.com/en-us /library/ms972976.aspx, accessed on December 20, 2015.
14. G. Smolkal. The OZ Programming Model. *EURO-PAR'95 Parallel Processing,* vol. 966, Berlin: Springer, 1995.
15. D. Grolaux, P. Van Roy, J. Vanderdonckt. QTk—A Mixed Declarative/Procedural Approach for Designing Executable User Interfaces, in *8th IFIP Working Conference on Engineering for Human-Computer Interaction* (EHCI'01), 2001, pp. 109–110.
16. Tcl Programming Language & Tk Graphical User Interface Toolkit. Available from: http://www.tcl.tk/, accessed on August 21, 2010.

Chapter 5

Human–Computer Interface in Pervasive Environments

Human–computer interaction (HCI) bridges the physical world and the digital world, providing communication channels for humans to send requests to and fetch responses from computer systems. Traditional HCI is restricted to a one–one mode where only one user interacts with one device (such as a personal computer) through a few communication channels such as a mouse and a keyboard. However, in past decades, technological development has diversified the devices with which humans interact (e.g., personal digital assistant [PDA], smartphone, tablet computer, etc.) and the modalities by which humans interact (e.g., speech, gesture, touch, etc.). For this reason, users now require HCIs to migrate their interactions across different devices or modalities in pervasive environments when they are not satisfied with current devices. To provide better user experiences under multidevice and multimodal environments, we propose an HCI service selection algorithm considering not only context information and user preferences but also interservice relations, such as relative location. Simulation results illustrate that the algorithm is effective and scalable for interaction service selection. We then propose a Web service–based HCI migration framework to support an open pervasive space full of dynamically emerging and disappearing devices.

5.1 Overview

Computer systems play a prime role in support of information delivery. Generally, the effectiveness of information systems largely relies on two factors: information quality and presentation. On the other hand, expansion of the Internet and other sources of digital media have provided people with access to a wealth of information. This trend (or *information overload*) features new requirements for both high information quality and smart presentation. Moreover, versatile sources of digital media greatly extend computer interfaces and thereby promote HCI technologies to a new generation.

Traditional HCI is restricted to a one–one mode where only one user interacts with one device, such as a personal computer, through a few communication channels, such as a mouse and a keyboard. Research on HCI has promoted new technologies and improvements in user experience. However, in past decades, the rise of mobile devices (e.g., PDA, smartphone, tablet computer, etc.) has impacted the monotonous interaction pattern between human and personal computers, and the development in computer vision and sound processing has enriched communication channels to digital media (e.g., speech, gesture, touch, etc.). Trends such as ubiquitous computing [1] gradually have made traditional HCI technology appear insufficient and inconvenient, while collaborative and distributed HCI technologies using multiple devices and interaction modalities are gaining more and more attention.

We offer two possible scenarios for multimodal and multiplatform HCI under a pervasive computing circumstance.

- *Scenario 1.* Suppose a user is taking a trip and bringing his camera. He takes photos of the landscape and would like to share them on a social network. Before uploading, he wants to edit these photos and select some of them for sharing. He then unlocks his mobile phone where the photos just taken are displayed. He edits some photos and uploads them to Facebook. While he is uploading the photos, he finds a new portrait of his friend. To obtain a larger view of the image, he steps near a public computer and that Web page is automatically displayed on the computer without any login requirements. In this scenario, interaction is migrated among the camera, mobile phone, and public computer.
- *Scenario 2.* Suppose a father is reading a business document on his personal computer. The document is displayed on the screen as a traditional PDF file. In a while, his 10-year-old son would like to use his father's computer to continue reading a story about Peter Pan he had not finished the night before. After the son takes a seat and opens the story, the computer automatically reads it to him aloud and displays it word by word according to the child's reading speed. In this scenario, interaction is migrated from a visual form to an audio form, according to different user contexts.

These two simple scenarios illustrate the following requirements for HCI within a pervasive environment:

- *One user, many devices.* Interaction needs to be migrated across multiplatforms according to context awareness such as physical positions of users and devices.

The whole interaction system must be able to dynamically detect nearby devices and establish appropriate connections with them. Moreover, interactions may be distributed among several devices.

■ *Many users, one device.* Interaction needs to be migrated across multimodals according to user preferences and contexts.

One solution to these requirements is called interaction migration, which is to migrate the interaction process among multiple devices, and even different modalities, in a graceful manner.

Interaction migration can be achieved at different levels. Process level migration is a straightforward intuition, but its tendency to form bottlenecks when transferring large data and reconstructing processes on various embedded platforms hampers its widespread use. Task level migration, on the other hand, extracts logic tasks and their relationships from an application using a model-based method [2]; then migration can be achieved by distributing tasks to multiple devices. However, although such migration techniques perform well at modeling logic and temporal relationships among tasks, to construct ontology and categorization of tasks is non-trivial, thus placing obstacles for migration among interaction modalities. In recent years, a more flexible and nature concept (or service-oriented concept) has been proposed for HCI modeling. The nature and progression of the service encounter has become a key concern of human–computer interface designers, which is similar to a majority of services supported and provided by commercial information systems. [3]. This service-oriented HCI concept transforms interaction migration problems to service selection problems and attempts to find solutions from Web service selection techniques.

We adopt the service-oriented concept and present a context-aware HCI service selection process that considers not only context information and user preferences but also interservice relations. Our main contributions are:

1. *Context-awareness method for service selection.* Take both interservice contexts and user contexts into consideration during migration request submission and service matching algorithm, making matching decision more flexible and intelligent.
2. *Service selection algorithm with good scalability.* Introduce interaction hot spot to evaluate service combination results.
3. *Formulate device matching during migration process as service combination discovery process.* Use service concept to describe interaction functionalities of application. This service-oriented concept generalizes interaction types and makes it possible to migrate interaction to multiple devices and modalities.

We further proposed a Web service–based HCI migration framework in which the interaction logic of application is modeled as an interaction service; through such modeling and ontology of services, it is possible to migrate interaction to multiple devices and modalities. Besides, user preference and context awareness are concerned in the framework by enclosing them into descriptions of interaction interfaces. Moreover, the framework supports dynamically emerging and disappearing devices in pervasive environments.

5.2 HCI Service and Interaction Migration

The proliferation of mobile devices and networks leads to an increasing demand for multidevice and multimodal interactions. The idea of interaction migration has been proposed to meet this demand in which HCI can be migrated across different platforms and modalities to provide better user experiences. Various interaction migration systems take advantage of model-based methods to separate interaction tasks and to generate corresponding interfaces on target platforms [2,4,5]. Some authors, however, use another mechanism based on Web service structure to provide migration under pervasive computing environment [1,6,7].

Our idea follows a service-oriented architecture (SOA) that models HCI by interaction services and then formulates devices' semantics matching as a service selection process. To address the issue of service selection or service combination discovery, researchers have developed language descriptions for Web services such as WSDL [8], BPEL [9], and DAML [10]. These languages successfully unify the structure and provide a formal description of Web service functionality. The shortfall of these languages is their ignorance of semantics. Therefore, ontology languages, such as OWL-S [11], are deployed to specify the Web service semantics.

Within these ontology-based service-matching methods [12–16], OWLS-MX [12] is the first hybrid OWL-S service matchmaker. It exploits features of both logic-based and information retrieval (IR)-based approximate matching. It uses IR to handle semantics, thus improving its performance in Web service matching [13], and uses a ranking method to decide the most suitable services. It also presents a device ontology that provides a general framework for device description. When the accuracy and scalability of semantic matching processes are considered, ranking surpasses other methods for its robustness. Other researchers take a look at a finer decomposition of applications. For example, ScudWare [14] successfully proposes a middleware platform for smart vehicle space. Different from other frameworks, it further divides service into many interdependent components. These components support migration and replication, thus making them able to be distributed onto different devices. The behaviors of components are crucial during the process of application migration and thereby are nontrivial to performing such partitions.

Besides service modeling, another important factor of concern in service selection is context. A context is "any information that can be used to characterize the situation of an entity. An entity is a person, place, or object that is considered relevant to the interaction between a user and an application, including the user and the application themselves" [17]. Combining contexts into service selection makes selection framework more flexible and more intelligent when meeting environment changes and user behaviors automatically [18] presents a context-based matching for Web service combination. The paper adopts an ontology-based categorization, a two-level mechanism for modeling, and a peer-to-peer matching architecture. In another implementation of context awareness [19], the authors add context attributes to the famous mobile service discovery system JINI. In another study [20], the authors highlight the context awareness in mobile network environments. They propose an algorithm for context-aware network selection.

Our approach also considers context in semantic matching. Our framework is similar to others [19] because we integrate context attributes into interaction device descriptions and use them to aid in service matching. In addition to context information, we also consider user preferences for each kind of service by recording interaction history. This kind of information helps match user-expecting devices with certain services, thus increasing matching accuracy.

In order to achieve better accuracy, we have made two assumptions: one is that services are well categorized so that different services are distinguished from each other; the other is that a powerful middleware is used so that services are described in a standard and unified format, which greatly reduces the difficulty for service matching. We believe that the two assumptions are reasonable. For the first, services in a pervasive computing environment can be categorized according to their corresponding hardware features because they are device-oriented and their service boundaries are definite. For the second, powerful middlewares are being well studied, and some of them are even OS level platforms. This progress in research and industry practices has made it possible to perform semantic matching on the middleware level.

In another study [21], the authors consider the problem of bottlenecks in centralized dynamic query optimization methods due to messages exchanged on a bad network. They present a decentralized method for the optimization processes. Another study has conducted interesting research on user measurement of the adoption of mobile services [22] and lists some of the constructs that may influence the user acceptance of services. Context, personal initiatives and characteristics, trust, perceived ease of use, perceived usefulness, and intention to use are mentioned as important constructs of their instrument.

Service selection is a classical problem in Web service–related research. There are several projects studying the problem of quality of service (QoS) empowered service selection. In one study [23], authors present a QoS-aware middleware supporting quality-driven Web service combination. They propose two service selection approaches for constructing composite services: local optimization and global planning [24], which offer similar approaches in service selection with QoS constraints in a global view. Both methods are based on linear programming and are best suitable for small-sized problems because their complexity increases exponentially when problem size increases [25]. The authors present an autonomic service-provisioning framework to establish QoS-assured end-to-end communication paths across independent domains. They model the domain composition and adaptation problem as classical k-multiconstrained optimal path (MCOP) problems. These works all use Internet-wide service selection and thus do not consider the service location during selection. In HCI, service location is an important factor in service selection. We use a local matching procedure to obtain the matching degree for each service and a global selection procedure to find the best service combination. Service combination is selected when a service can effectively cover some areas, defined as interaction hot spots, where a user can effectively meet the HCI requirements.

In another study [26], the authors introduce a context-based collaborative selection of SOA services. Although it takes service locations into account, the distance-based location is not sufficient for HCI service selection because most devices/services not only have an interaction range requirement but also the best interaction angle. In two more studies [27,28], the authors give an introduction to HCI migration. They mention the problem of device location but do not define location constraints.

Based on the idea of interaction service, we further design a framework using Web service technologies to support HCI migration in environments composed of dynamically emerging and disappearing devices.

Previous works have designed several frameworks for multiplatform collaborative Web browsing. A good example of collaborative Web browsing can be found in WebSplitter [29], a framework based on XML splitting Web pages and delivering appropriate partial views to different users. Bandelloni and Paternò [30], adopting similar ideas, implemented Web pages migration between a PC and a PDA by recording system states and creating pages with XHTML. This research provides access for users to browsing Web pages on various types of devices. However, the framework is only for collaborative Web browsing.

The main obstacle to designing a framework for general application user interfaces is the heterogeneity of different platforms. With various limitations and functionalities, it is hard to decide the suitable UI elements regarding the target platforms. One model-based approach [5] tried to define application tasks and modeled temporal operators between tasks. This ConcurTask Trees (CTT) model divides the whole interaction process into tasks and describes the logic structure among them, which makes the responsibilities of target platform user interfaces clear. Patern'o and Santoro made great contributions within this realm and offered detailed support for applying the model [31,32]. Another important issue is automatic generation of user interfaces. Unfortunately, automatic generation is not a general solution due to the varying factors that have to be considered within the design process, although a semiautomatic mechanism is more general and flexible [33].

Based on a CTT model and XML-based language, TERESA [34] supports transformation from task models to abstract user interfaces and then to user interfaces on specific platforms. The principle is to extract an abstract description of the user interface on a source platform by a task model and then to design a specific user interface by the environment on the target platform. TERESA is able to recognize suitable platforms for each task set and to describe the user interfaces using an XML-based language, which is finally parsed along with platform-dependent information to generate user interfaces. TERESA tools perform rapidly in deploying applications that can be migrated over multiple devices, but it addresses only the diversity of multiple platforms while neglecting the task semantics of users. With these insights, Bandelloni and Paternò [35] integrated TERESA to develop a service to support runtime migration for Web applications. However, the application migration service designed by Bandelloni needs a centralized server and device registration, which is costly and unadoptable for new devices. Moreover, due to its being based on task model, which is constrained by the task functionality, it is unable to handle multimodal interactions such as visual to audio.

5.3 Context-Driven HCI Service Selection

The basic idea of our design is that, when encountering interaction migration, users are usually inexpert in deciding the target platform because they do not have the insight into device capability and compatibility. Therefore, our selection method aims at providing an automatic selection mechanism rather than a user-initiative selection with the help of context information and user preference. We also separate interaction tasks from application logic and model them as *interaction services*. With the help of service description technology, we propose a structure for *interaction service selection* to achieve multimodal and multiplatform migration.

5.3.1 Interaction Service Selection Overview

Figure 5.1 shows an overall description of the selection process in our framework. Three main components comprise the interaction environment: the *user device*, which contains user identification and personal information; the *context manager*, which maintains context information caught by various sensors; and the *interaction device*, which performs service matching and provides interaction interfaces. The following steps are involved in the whole process.

1. *Initiate migration request.* Migration requests can either be driven by context or announced by users. In both situations, those requests are sent to the context manager. In our design, the context manager takes charge of all sensor networks in the environment, while the user device holds the user identification. The context manager is able to trace the changing contexts of a certain user by monitoring parameters such as user position, facial expressions, and gestures. A key idea is that it is often natural for a user to migrate interaction when context changes. For instance, a user may express the need for

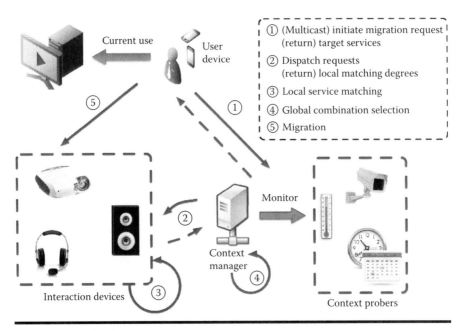

Current use
User device

① (Multicast) initiate migration request (return) target services

② Dispatch requests (return) local matching degrees

③ Local service matching

④ Global combination selection

⑤ Migration

Monitor

Context manager

Interaction devices

Context probers

Figure 5.1   Selection process overview.

interaction migration by performing a certain gesture or by simply approaching target devices. Techniques such as human behavior modeling [36] are potential solutions for such context detection. Therefore, for a user-driven migration request, the user device directly forwards the migration request to the context manager. Notably, although initiated by the user, the migration requests do not directly specific target devices; this is different from a user-initiative selection method. On the other hand, when a switch in context is detected (context-driven situation), the context manager will notify the user device to forward the migration request. This seemingly redundant behavior is necessary because the context manager does not hold any user information (for privacy reasons).

2. *Dispatch requests.* The whole service selection process is divided into two parts: local service matching and global combination selection. Local service matching is performed on device terminals in a distributed manner. Therefore, after receiving migration requests from a user device, the context manager dispatches requests to devices near the user for local service matching. The degree of matching will be sent back to the context manager after calculation. Future improvements such as device filtering and request reforming can be implemented within this step to address accuracy and performance concerns.

3. *Local service matching.* Local service matching is performed on device terminals and returns the degree of matching to the context manager.

Service functionality requirements, user preferences, and contexts are taken into account in the algorithm. Those interaction devices that cannot provide these services will not respond to the request, thus alleviating burdens on network transmission within a device-rich environment. Context information can be retrieved from the context manager if needed in local selection matching.

4. *Global combination selection.* After receiving the matching degrees calculated by device terminals, the context manager selects the best service combination for the user's requests. User request, interservice contexts, and local matching degrees are integrated here to come up with the best service combination. Due to the complexity of the combination problem, we provide an approximate algorithm for finding the optimal service combination. The final selected service combination results are returned to the user device.

5. *Migration.* After the user device obtains target interaction devices, it multicasts connection request to those targets. Selected devices then send back Web service descriptions so that connections between the user device and the targets can be established. Subsequent interaction processes are not included in this section because we focus on service selection.

5.3.2 User Devices

5.3.2.1 Service-Oriented Middleware Support

Although having been proposed for decades, the concept of interaction migration has not been implemented in practical applications. The major obstacle is that it is hard for application developers to cover tedious details of interaction processes and to modify existing applications to support such migration. To solve this problem, we propose an HCI migration support environment (MSE, refer to Section 5.5), which exists between the application level and OS level, acting as middleware, hiding platform-dependent details from users, and providing interaction migration Application Program Interfaces (APIs) for upper applications. Here, we first assume there is middleware in which HCI processes can be fulfilled by calling interaction services. In the system, local and remote interaction services provided by interaction devices can be called with a unified interface. Therefore, when encountering interaction logic, the upper level applications just call APIs provided by the middleware and let it select target interaction services. Hence, interaction migration can be treated as seeking proper remote service combinations that can fulfill requested interaction requirements. The service-oriented middleware then needs to take charge of extracting service descriptions based on an application's interaction behavior. Two questions should be answered in the descriptions: "what services are needed?" and "what are the basic properties of these services?" The system can answer the two questions by tracing the APIs invoked by upper level applications. We categorize APIs into several interaction service types (such as *video display service* and

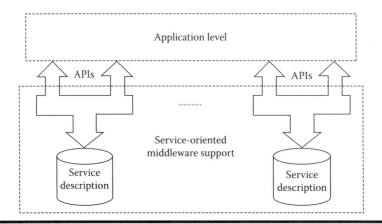

Figure 5.2   Service-oriented middleware support.

---

keyboard input service) and record key parameters as service properties. Therefore, once an application calls an API, our middleware can specify corresponding types and properties of services called by the application (shown in Figure 5.2).

5.3.2.2 User History and Preference

We assume that different users possess different preferences for interaction devices, and such preferences can be acquired from users' interaction histories. For example, a user who prefers large screens may have a higher average screen size in his previously used devices. Therefore, catching the parameters of a user-expecting device from a history is crucial in selecting appropriate interaction devices. In our method, we adopt a service-oriented concept and associate parameters of device hardware with interaction services. In addition, the interaction services are categorized into independent groups such as video output and keyboard input to facilitate service matching.

Assume that there are k service categories $C = \{c_1, c_2, \ldots, c_k\}$. Let a device d with hardware feature set $F_d = \{f_1, f_2, \ldots, f_n\}$. We assume that it can provide services $S_d = \{s_1, s_2, \ldots, s_m\}$. The properties P_i of service s_i are a subset of the hardware feature set; that is, $P_i \subseteq F_d$. We then extract general properties from set service properties, since each service can be assigned to a service category. General properties actually belong to a service category because they are used by all services in the same category. For instance, a desktop computer commonly has two service categories: *video output* and *audio output*. If it has two displays, then there are two interaction services on the computer that both belong to the video output category. *Screen size* and *resolution* are general properties of these two services.

In user devices, we maintain a service property set U^c to represent user preferences and to describe a user's expected properties for service category $c \in C$. Notably, we ONLY record *general properties* for a service category because only those common properties can be used for comparisons among the services in a category. User preferences are formed from device-using records in the past history. After a user selects a service *s* from category *c*, the user device will record the general properties of the service and update the *i*th properties of corresponding user preferences by:

$$u_{i,k}^{c} = (1 - \alpha)u_{i,k-1}^{c} + \alpha c_{i,k} \tag{5.1}$$

where $u_{i,k}^{c}$ denotes the *i*th property in user preferences for category *c* after using a service from *c* for the *k*th time; α denotes the update rate for the newly incoming device descriptions; and $c_{i,k}$ denotes the *i*th general property value of the *k*th time selected service. Suppose the initial user preference is $u_{i,0}^{c}$, then a candidate value for α is $1/k$, which means the user preferences are calculated by averaging past device parameters. When requesting migration to some services, a user device will envelope user preferences for these services and send them to nearby devices for service matching.

5.3.3 Context Manager

Our goal is to propose a context-aware method for selecting proper interaction services that makes it necessary to deploy a context manager to monitor the environment and users, to record and analyze contexts, and to provide global access for interaction devices. Another important responsibility of the context manager is to integrate multimodal context information such as speech, gestures, writings, facial expressions, or combinations thereof. The context manager runs a software system connected with user devices and sensors. We do not put much emphasis on how the context manager connects with sensors although we do provide insight into what contexts should be provided and how we use the information. We assume that the context manager is able to monitor data from sensors and to detect context switches along with user behaviors. The context manager should maintain two major types of context information:

■ *Environment context.* Including environment temperature, moisture, brightness, current time, and so on, environment context is crucial to those interaction devices that have some running constraints under environmental condition. For example, high temperature and moisture may greatly influence some sensitive devices, decreasing their process ability and thus harming user experiences. Another interesting use of environment context is related to time records. For instance, a device may record when its services are being used, compare the current time with the records, and judge whether it is appropriate for providing such services.

■ *User context.* Researchers have made great progress in HCI technology [37], such as face detection [38], expression analysis [39], gesture and large-scale body movement recognition [40], and even eye tracking [41]. For a smart pervasive environment, it is usual to deploy these technologies, corresponding sensors, and algorithms; these are called *context probers* in our framework. Our context manager monitors these context probers and fetches context information from them. A simple example is how user position matched to the nearest interaction devices. We can also apply user face orientation information based on detection technology to match proper screens in a user's vision.

In our method, context information mainly aids interaction migration in two ways. First, context information helps determine the moment when to launch interaction migrations. Second, a comparison between contexts and device parameters improves service matching accuracy. Moreover, the context manager is responsible for the algorithm of service combination selection that will be discussed in the following section.

5.3.4 Local Service Matching

Local service matching in Figure 5.1 is performed on each interaction device that receives migration requests from the context manager. Matching degrees are then sent back to the context manager for further selection of service combinations. Interaction devices are designed to calculate matching degrees because the context manager will be a network bottleneck if too much device and service information needs to be transmitted back. A device profile is maintained on each interaction device to compare with service descriptions and with contexts. The service matching procedure contains the following three parts:

- *Service property matching.* The top priority is to judge whether the current interaction device is capable of providing the service required; that is, to determine whether the device can meet the basic properties of required service. Our solution is to keep device capability information in a device profile, including supported service types and properties for each service type. Hence, the interaction device merely needs to judge whether requested services exist in supported service categories and whether corresponding service properties are within the capabilities of the HCI service. If service property matching fails, the device will not reply to the user device. Otherwise, the following two matching processes will be performed, and a matching degree will be returned.
- *User preference matching.* Our goal is to select what are potentially the most satisfying services for users. Therefore, this part calculates to what extent the interaction device satisfies a user's expectation—namely user preference—when considering certain services. User preferences are enclosed in the migration request, although device features for its services are recorded in a device profile.

■ *Context matching.* The main contribution of our method is the utilization of context. We first retrieve context information from the context manager, and then we evaluate environment and user context information by a context evaluation function, preset in interaction devices, to measure whether devices are suitable to provide services under current circumstances. The context evaluation function $f_c(x)$ for a context c varies according to the actual semantics of the context. For example, *Euclidean distance*

$$f_{\text{position}}(x) = \left\| \vec{x} - \vec{p} \right\|_2 \tag{5.2}$$

where p denotes the position of a device, performs well for position context to measure whether a user is close to the device, while *cosine similarity* metric

$$f_{\text{cosine}}(x) = \frac{\vec{x} \cdot \vec{o}}{\left\| \vec{x} \right\|_2^2 \cdot \left\| \vec{o} \right\|_2^2} \tag{5.3}$$

where o denotes the orientation of a device, is suitable for context variables such as user face orientation and screen orientation. Interaction devices can also specify weight w_{d,s_i}^c for each context c to denote how significant the context will influence service s_i's selection on device d.

The final matching degree is summed up from the results of user preference matching and context matching. Matching degrees are calculated for each requested service. After collecting matching degrees, the context manager starts a global service combination selection.

5.3.5 Global Combination Selection

We have presented a local service matching procedure by which matching degrees can be derived for each requested service. In simple scenarios, the information from service matching degrees is enough for a successful interaction migration. However, there are limitations that may lead to unreasonable migration. For example, because the local service matching procedure considers little about the interservice contexts such as relative distance between two target devices, it may result in a bad user experience if the distance is too long. Therefore, we propose a global selection procedure to find the best service combination. We focus on the relative locations as an example of processing interrelations among services. In fact, more general interservice relations can be modeled in our framework using similar approaches. In order to represent the relationship among different services in terms of relative distance, we develop a service-coverage model and corresponding search algorithm. The *coverage* idea is similar to target coverage scheduling proposed elsewhere [42]. Each interaction service in our framework resembles directional sensors in Han et al. [42], but we use soft evaluation rather than hard classification for our effective region of interaction services.

5.3.5.1 Effective Region

In reality, different services on different devices have different ranges to provide effective interaction service. We use effective region to represent the range within which a user can interact with a service under a QoS guarantee. The effective region for a device's interaction service without any physical barrier can be reasonably modeled as a sector centered at the device. The sector angle represents the device's interaction angle. As illustrated in Figure 5.3, a slashed area denotes the computer's effective region within which users can use the computer effectively.

For a given device, different services may have different effective regions. For example, visual service may have a smaller effective region than audio service. Even for a given device's interaction service, HCI effectiveness may vary from point to point within its effective region. In Figure 5.3, users may feel it is more convenient to use the computer at point A than at point B. Therefore, we use function EV_i for service i to represent the HCI effectiveness at point p:

$$EV_i(p) = \begin{cases} 0, & \text{if } p \text{ is out of effective region} \\ (0,1], & \text{otherwise} \end{cases} \tag{5.4}$$

The effectiveness can be zero if p is out of one device's effective region while within the effective region of another device, the function returns a device- and service-dependent score between $(0,1]$.

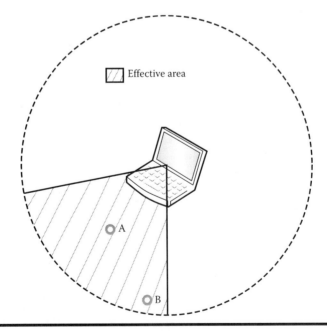

Figure 5.3 The effective area is modeled as a sector.

5.3.5.2 User Active Scope

Interaction migration happens only when the user enters a different area that is called the user active scope (UAS). The UAS information is maintained by the context manager. We assume that the context manager can properly divide user space into different UASs according to the information about building inside a structure that can be initially set in the system. Meanwhile, users are inclined to move within a UAS for effective interactions.

It is flexible to define a UAS that depends on both environment and applications. For example, a room can be a single UAS in some situations while, in other applications, it may be divided into several UASs.

In Figure 5.4, a UAS is covered by several continuous squares. A UAS square is an atomic unit that users can step in or out of to interact with some devices. The UAS square will be used in the service combination selection algorithm we present later. It is obvious that a large UAS square can speed up our selection algorithm, but it will decrease its accuracy.

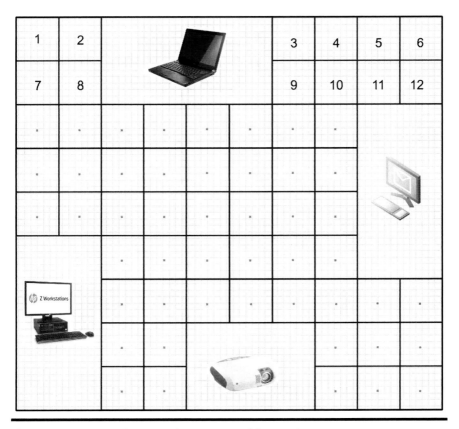

Figure 5.4 *User active scope* **is represented by continuous squares.**

5.3.5.3 Service Combination Selection Algorithm

Service location is concerned with HCI migration. Actually, two individually best matched services usually cannot be combined to provide HCI services in one application if they are not close enough. Therefore, we develop a global combination selection algorithm to take service location into account. Our algorithm includes the following two procedures:

1. *Service-coverage coloring procedure.* Suppose there are M different services in a UAS and, for each square unit, we use an M-bit string to represent whether users can effectively interact with the corresponding service in the squares. If the effective region of service i covers square p, then $sc[p][i]$ is set to 1; otherwise, 0. We call this step *coloring*, and it is performed on the context manager. The coloring procedure running on the context manager includes following three steps:

   a. The context manager retrieves information about a user's current UAS.

   b. Divides the UAS into N continuous squares.

   c. For each of the M services, the context manager colors the squares within its effective region.

   Figure 5.5 shows the result of the coloring procedure for the UAS in Figure 5.4. For a given UAS, the coloring procedure only needs to run once after its device position changes.

2. *Service combination selection procedure:* Suppose there are K categories of services required to be migrated. Different devices can provide the same category of services. So we need to find out the best service for each required service category. We denote services in current UAS as set S and the service category set as C. Therefore, we have $\forall c \in C$, $c \subseteq S$ and $\forall s \in S$, $\exists c \in C$, s.t. $s \in c$. A subset S' denotes those returned (matching) services. Denote the service category set of S' as C'. Then $\forall c' \in C'$, $c' \subseteq S'$ and $\forall s' \in S'$, $\exists c' \in C'$, s.t. $s' \in c'$. Matching degree d_s, $s \in S'$ denotes a local matching degree for service s. Denote position (square) set as P. Then the coverage information for service s and position p can be denoted as $sc[p][s]$, $p \in P$, $s \in S$. We now suppose the context manager has derived matching degrees for the current migration through a local service matching procedure. The context manager then performs as follows:

   a. For each square p ($p \in P$), find the most suitable service i for service category c satisfies:

   $$s_p[c] = \max\{EV_i(p) * d_i\}$$
   $$\text{s.t. } sc[p][i] = 1, \ i \in S', \ i \in c, \ c \in C' \tag{5.5}$$

   b. Calculate the overall matching degree for square p by:

   $$t(p) = \sum_{c \in C'} \left(w_c * S_p[c] \right) \tag{5.6}$$

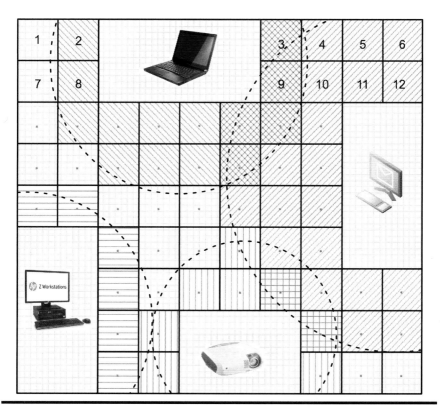

Figure 5.5 Results of the coloring procedure are represented by continuous squares.

where $t(p)$ is the overall matching degree for square p and w_c is the weight of the service category c. w_c is determined by applications.

c. Select the square *sel* with the maximum overall matching degree:

$$sel = \text{argmax}\{t(p)\} \qquad (5.7)$$

The square *sel* will be the best position where user can have most effective interaction, and the corresponding services will compose the best service combination.

5.4 Scenario Study: Video Calls at a Smart Office

In this section, we study a specific scenario, having video calls with a customer at a smart office, to illustrate how our context-aware service selection framework works to provide better user experiences.

5.4.1 Scenario Description

5.4.1.1 The Smart Office Environment

As smart devices become more and more common in our daily life, the concept of the *smart office* becomes a hot idea in pervasive computing. Thus, we start our scenario study in such a typical smart environment.

One smart office may consist of the following devices:

1. Displays, projectors, and so on, categorized as visual devices because they can output visual images.
2. Loudspeakers, wireless headphones, and forth, categorized as audio devices because they can output audios.
3. Microphones, categorized as voice devices, for the reason that they can input voices.
4. Video cameras, categorized as video input devices.
5. Desktop computers, laptops, and tablet computers, categorized as compound devices because they can provide multiple services.
6. Printers and other irrelative kinds of devices.

Figure 5.6 shows the layout of our smart office. There are four working blocks in the office with a desktop computer in each block. Computer 01 has an embedded

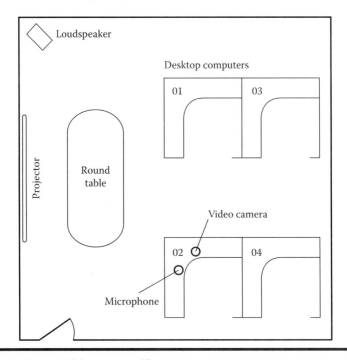

Figure 5.6   Layout of the smart office.

video camera and microphone. Near Computer 02, there is an external video camera and a wireless microphone. The devices and corresponding services contained are listed in Table 5.1.

To construct this smart environment, these devices must satisfy two basic requirements. First, they themselves must be smart, which means that certain computing ability is required so that they can *think* independently. Second, they must be connected together, either in a wired or wireless way, so that they can communicate with each other in order to act together. Development in mobile and wireless networks has presented several approaches for connecting these devices [43].

A smart office should also be aware of the status of both users and the environment. Thus, sensors become its most important part. For example, heat sensors can be used to collect information about temperatures; surveillance cameras can be used to track users' motions and their face orientations. The data collected by these sensors are called context information.

Table 5.1   Devices and Services Used in Simulation

Device Name	Service Name	Category
Projector	projector_display	VO
Loudspeaker	loudspeaker_sound	AO
Computer 01	comp01_display	VO
	comp01_sound	AO
	comp01_camera	VI
	comp01_micro	AI
Computer 02	comp02_display	VO
	comp02_sound	AO
HD VideoCam	ext_video_camera	VI
Wireless Micro	microphone	AI
Computer 03	comp03_display	VO
	comp03_sound	AO
Computer 04	comp04_display	VO
	comp04_sound	AO

Note: AI, audio input; AO, audio output; VI, video input; VO, video output.

5.4.1.2 Scenario Description

Suppose there is a user, named Bob, having a video call on his smartphone. Due to the small screen of his phone, he may not be satisfied. So he walks into the office that has better devices. In such a scenario, our framework should be able to detect Bob's intentions, find the devices that can better satisfy Bob's need, and migrate interaction to these new devices.

5.4.2 HCI Migration Request

An *HCI migration request* is sent by a user device. A user device is a core device in our framework. It can be used to identify the user and to decide whether to start an application migration when the user's context is changing. One typical user device may be a smartphone, for the reason that it is portable, easy to obtain user preferences, and has good process ability. People frequently use them to check e-mails, surf the Internet, and to connect with different devices. All these activities imply users' preferences toward certain services and devices, which can be recorded as *user preferences* by the smartphone in order to facilitate service matching.

After the user device decides to start an HCI migration, requests are sent to the context manager and then dispatched to device terminals. The request is written basically in XML format. It is made up of three parts: the service description, user preference, and user identification information. The service description should contain descriptions of an application's related services. These descriptions must contain the essential requirements, namely service properties, so that receivers can decide whether they can offer such services based on these requirements. User preference contains information for devices to decide whether their services can meet the user's need. Finally, user identification information is provided to identify who initiates the request.

In this scenario, because Bob is having a video call on his smartphone, which is powerless to present videos and sound tracks, *video output service, audio output service, video input service,* and *audio input service* are four important service categories to be contained in the migration request. Bob's requirements can be explained as the migration of these four services in our framework. Figure 5.7 illustrates the structure of the migration request within our scenario. Among the four service categories, *video output service* is investigated the most. Three properties—format, resolution, and bandwidth—denote the supporting video format, video resolution, and transmission bandwidth, respectively. An XML format of this request can be found in Appendix 5.7.1.

5.4.3 Context Format

As is mentioned earlier, the context environment is made up of the data collected from different sensors. These data are stored on a local server called the context

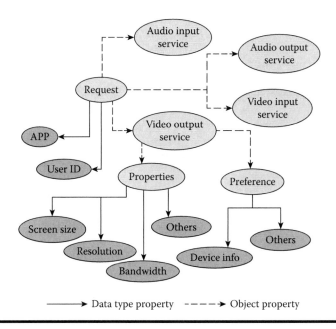

Figure 5.7   Structure of service migration request in video call scenario.

manager, which is responsible for monitoring context information and for providing access for interaction devices. Context information consists of two parts: environment and user context. In our scenario, several user contexts need to be considered by our framework (such as user position and face orientation) because, within these contexts, those devices with video display service having screens that are close or nearly face to face with the user will achieve a high degree of matching in service matching. The structure of context information retrieved from the context manager is relatively simple (as shown in Figure 5.8). An XML format is given in Appendix 5.7.2.

5.4.4 Device Profile

The device profile is rather essential to the local service-matching algorithm, as is mentioned in Section 5.3.4. Each device has their own device description (the *device profile*) that describes their abilities. The profile contains several service descriptions, each of which is used to describe one of the device's features when offering this service. Similar to the format of a request, the service description in a device profile also contains corresponding parts for matching service properties and user preferences, as is mentioned in Section 5.3.2. What seems different is that context information is considered here. The relevant context information may vary with different devices and services. For example, in our scenario, Computer 02 can receive input from a keyboard, display videos, and can play sound tracks.

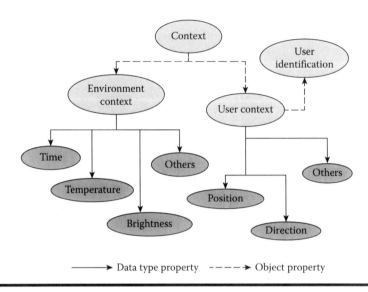

Figure 5.8   Structure of context information in video call scenario.

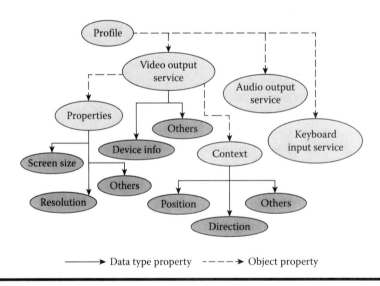

Figure 5.9   Structure of device profile of Computer 02 in video call scenario.

Then these three abilities are categorized into three different services: *keyboard input service*, *video output service*, and *audio output service*. Figure 5.9 illustrates the schema structure of Computer 02's device profile, and some essential properties and contexts of video output service are illustrated.

In our scenario, when Computer 02 receives a migration request and context information from the context manager, it then starts a local service matching process.

It first compares the service name and properties between the request and its device profile to know whether it can offer this service. The result is that video output service and audio play service are within the matching scope; keyboard input service is unused; and video input service and audio input service are not supported. Second, for these successfully matched services, device information regarding these services will be compared with user preferences in the request to measure the divergence to the user-expecting device. Finally, Computer 02 considers context information by calculating distance to the user and cosine similarity between screen and user face orientations. The final matching degree is then sent back to the context manager. After receiving matching degrees, the context manager will perform a global service combination selection. In the following section, we illustrate the simulation result of this process for our scenario.

5.4.5 Experiments and Results

We illustrate simulation results of our context-aware HCI service selection algorithm applying to the smart office scenario discussed earlier and compare our results with the method presented in Wang et al. [44] and show that modeling context information and user preferences alone is insufficient for discovering an optimal service combination. Our service selection algorithm that considers interrelations among services is much more effective in searching target service combinations for interaction migration. As is mentioned earlier, we categorize interaction services into several groups and pick the best matches in each group to form the best service combination regarding a position unit. In this simulation, we predefined four service categories: *video output service, video input service, audio output service,* and *audio input service.* Each device in the simulation contains several services from one or more service categories. Services provided by different devices may consist of various descriptions or *EV* functions due to the diversity in device parameters. It is easy to extend this simulation to more complex situations by simply adding customized service descriptions and registering through a uniform service interface. The devices and corresponding services contained within are illustrated in Table 5.1.

5.4.5.1 Simulation Result

Our simulation aims at giving solutions to the smart office scenario mentioned in Section 5.4. Our service selection algorithm first scans each continuous square in a user's UAS and finds the best service combination. Figure 5.10 shows overall matching degrees for Equation 5.6 for each square in the room with regard to different selected services. The room setup can be referred in Figure 5.6. The room is divided into many squares (which are not shown due to the large number), and each square is colored by #000000 to #FFFFFF corresponding to the matching degree in that square. A deeper color means a higher matching degree in that square regarding

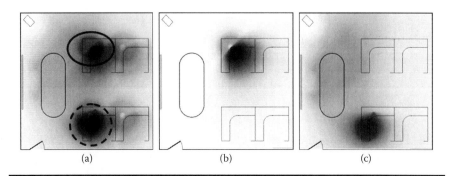

Figure 5.10 Overall matching degrees (Equation 5.6) of each square in the room with regard to different selected services. (a) matching degrees for the local optimal service combination of each square in the room; (b) matching degrees of each square for the globally optimal service combination; (c) matching degree of our service selection algorithm.

the selected services and therefore a better user experience with the selected services in that square. The matching degrees are normalized to [0, 1]. Figure 5.10a shows matching degrees for the local optimal service combination of each square in the room. The graph shows two potential areas for interaction migration—marked by solid and dashed circles in Figure 5.10a. The best service combination selected in the solid circle is {comp01_display, comp01_sound, comp01_camera, comp01_micro}, and the best one in the dashed circle is {comp02_display, comp02_sound, ext_video_camera, microphone}. It is reasonable to have multiple high scoring areas in a UAS because different service combinations may satisfy a user's migration requests. For the final decision of interaction migration, one may list all potential service combinations in matching degree order, or one may just pick the service combination with the highest score.

After assigning local matching degrees to each square, the best service combination is then picked from the square with the highest matching degrees (global selection procedure). In our simulation, the squares with the best matching degrees occur near the device Computer 01 whose service combination is {comp01_display, comp01_sound, comp01_camera, comp01_micro}. The target device for interaction migration is then Computer 01. Figure 5.10b illustrates the matching degrees of each square for the globally optimal service combination. One observation is that the chosen services are close to each other (actually in the same device). This is an ideal situation for interaction migration because users can easily have access to devices thereby increasing user experience.

For algorithm comparison, we apply the service selection algorithm in another study [44] to this smart office scenario and present the result in Figure 5.10c. The selected services in Figure 5.10c are {projector_display, loudspeaker_sound, ext_video_camera, microphone} and matching degrees are highlighted with gray density. Each individual service selected is the best one that matches the user's requests (e.g., the selected device *projector* is powerful in video output due to its large screen and high resolution). However, the overall matching degree in the *room* of the selected service combination is much less than what we selected (Figure 5.10b). The reason is our algorithm takes interrelation of services into consideration that benefit from discovering a good service combination. The room setup can be referred in Figure 5.6. The room is divided into many squares (not shown due to the large number) and each square is colored by #000000 to #FFFFFF, corresponding to the matching degree in that square. A deeper color means a higher matching degree. The matching degrees are normalized to [0, 1]. (a) Matching degrees for the local optimal service combination of each square in the room. (b) Matching degrees of each square for the globally optimal service combination. (c) Matching degrees of each square using the selection algorithm [44].

5.4.5.2 Scalability

Response latency is always a crucial factor in HCI technology. For service-oriented interaction migration, selection algorithm may be a bottleneck for the overall migration process. An efficient service selection algorithm needs to be scalable with an increasing number of services. In our algorithm, we compute matching degrees for each square divided in UAS by selecting the best services within each service category. Let the number of service categories be C, each service category contains S services, and the total number of squares divided in UAS be N. The overall complexity of our service selection algorithm is $O(CSN)$. The computing time grows quadratically on average as more services are added because C is linearly related to S. Another key factor influencing computation time is the number of squares. For a fixed size interaction migration area, more squares that are divided usually means more accurate results for service selection but more computing costs. Fortunately, the computing time grows linearly with the number of squares (as shown in Figure 5.11). Moreover, the area of a square usually is not necessarily too small because users may feel no difference between two different squares if they are smaller than the users' sensitivity criterion.

5.5 A Web Service–Based HCI Migration Framework

We propose a Web service–based HCI migration framework to support open pervasive space full of dynamically emerging and disappearing devices based on the idea of the aforementioned HCI service selection. To avoid costly centralized services and improve scalability, our framework resides on each device, composing a peer-to-peer structure. With user preferences and context awareness implemented in each device, we do not need a centralized context manager. Therefore, there will be some differences between the HCI migration framework and the service selection structure presented earlier, including system components and interaction processes.

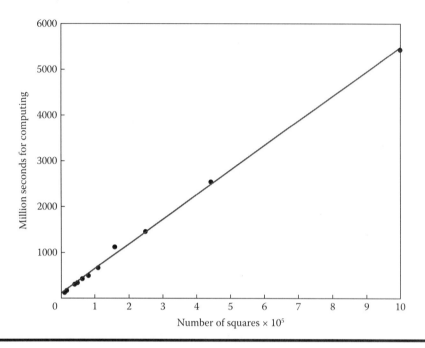

Figure 5.11 Computing time for increasing number of squares divided with a fixed number of services.

5.5.1 HCI Migration Support Environment

Traditional implementation of application user interface migration relies on centralized service, which performs costly and awkwardly when faced with newly emerging devices. Static device registration is incompetent in open pervasive surroundings because, contrary to a local smart zone, we can never know what devices will be involved in an interaction area. Today, however, users tend to prefer services that can support dynamic updates of device information; otherwise, they instead would use one device to finish the work regardless of its inconvenience [45]. Therefore, our framework design for interaction migration aims at the *hot-swapping* of interaction devices. As shown in Figure 5.12, the HCI MSE lies between the application level and OS level, acting as a middleware, hiding platform-dependent details from users and providing interaction migration APIs for upper applications. For each platform within the interaction zone, an MSE defines the boundary of application and makes the application free from the burdensome affairs of multiplatform interaction creations such as communication with other MSEs and so on. The application merely needs to submit interaction requirements and tasks and then leave the details to the MSE. With each device represented as an *interaction Web service* (discussed in following sections), the whole connection structure resembles a peer-to-peer framework and thus avoids centralized services.

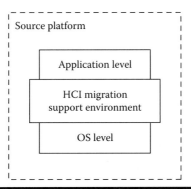

Figure 5.12 Framework position between application and OS system.

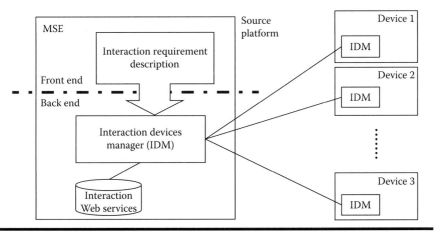

Figure 5.13 Framework of HCI migration support environment.

The framework can be divided into two ends (shown in Figure 5.13):

1. A *front end* that is platform-independent and responsible for handling user preferences and context awareness.
2. A *back end* that is platform-dependent and in charge of interaction migration and communication with other devices.

The front end formalizes user preferences, context information, and application constraints as the *interaction requirements description*. The back end then utilizes these descriptions as input for the *interaction devices manager* (IDM) that fetches device information from the network and exerts semantic matching to select appropriate devices for migration. The IDM also maintains platform-dependent information and acts as an interaction device to respond to requests from other platforms. Devices that provide interaction are represented as *interaction Web services* and are provided for semantic matching. After selecting target devices, interactions are performed through networks.

5.5.2 Interaction Requirements Description

HCI involves both human behaviors and application manners. With this consideration, we need to formalize a description, namely an interaction requirements description (IRD), to include both user requirements and application response.

Figure 5.14 demonstrates the composition of an application in general: *program logic*, *interaction elements*, and *interaction constraints*. Program logic assures the correct reaction of an application to certain interaction input. Therefore, program logic can be separated from the interaction framework, which makes our framework independent of application functionality. The IRD includes three parts:

1. *Definition of interaction elements.* For graphical user interface (GUI), interaction elements include buttons, links, and other interface elements. In our condition, which is more general, the definition covers all input and output items during interaction and the descriptions of their names, types, and functionality.
2. *Interaction constraints.* There are two sources of constraints. Requirements on interaction devices proposed by users and context-driven constraints are formalized as constraint descriptions used to select proper interaction devices. For example, the physical position relationship between a user and devices is taken into consideration. The second kind of constraint is involved in the application interaction process. For example, if application interface contains audio material, constraints should involve *support for audio output.*

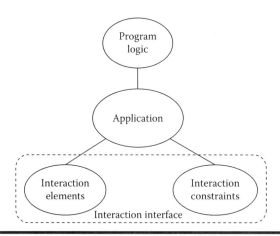

Figure 5.14 General compositions of applications.

When conflict occurs between user-defined and application-dependent constraints, we grant user-defined constraints a higher privilege because interactions need to more willingly satisfy users' feelings.

3. *Interaction display moment.* Defined by program logic, the moment for interaction display assures the continuity of the application process. When reaching the moment for interaction display, the application sends interaction requests through the MSE. All the elements displayed at the same moment make up the distributed HCI.

5.5.3 Interaction Devices Manager

Heterogeneous device structures always tax multiplatform interaction because it is hard to design a universally compatible interface, and it is costly to maintain several copies for multiplatform support. Another crucial problem for interaction migration is the matching of appropriate platforms. An intuitive idea is to analyze the migration requests from application users and fetch the most fit device service according to device description. Fortunately, the IRD discussed earlier has combined user requests, physical contexts, and application interaction elements, contributing to the accurate and swift matching of interaction devices. Nevertheless, there still remain problems for managing device information and providing a proper matching mechanism.

Our solution is to deploy an IDM to take charge of each connected device. When a new device emerges, a new instance is created by IDM to represent the corresponding device in each device within the network connection. The instance is a *skeleton* representation of devices to reduce the storage burden, from which contents are fetched when applications request migration. When receiving migration requests, the IDM first creates an IRD based on applications, user behaviors, and context, and then sends an IRD to connected devices. Each remote device is represented as an *interaction Web service*. Based on this concept, an IDM on each remote device maintains an interaction Web service description (IWSD), which includes interaction-related information such as display resolution, window size, and supports for other interaction modes. Semantic analysis is then carried out to calculate the matching degree between its own IWSD and the provided IRD. Semantic similarity measures the ability of the devices to fulfill the constraints of this interaction. All matching degrees are returned to the source platform for a final decision. For a successfully matched device (the one with the highest similarity), MSE announces these devices and locks the relevant resources for this device to avoid its being occupied by other interaction requests. It is entirely possible to match multiple devices (devices of the same kinds), and an MSE on a source platform will decide according to interaction semantics whether to distribute parts, or the total process of interaction, to those devices. Once interaction devices are decided, migration requests will be sent to device terminals and a connection will be established.

5.5.4 Interaction Web Service

The best feature of our framework structure is the absence of a centralized server, which is replaced by an HCI MSE on each device. Such framework is similar to the well-known peer-to-peer structure. Those devices that receive interaction requests from an MSE on a source platform are represented as IWS for application users. For each IWS, an IDM takes charge of maintaining the IWSD of the device (by static configuration or auto-generation), which is then used to compare it with a received IRD from a source platform. The matching degrees are returned to the source platform. The MSE on the source platform selects devices with the most fitting Web service description to the IRD and decides target platforms (see Figure 5.15). The key consideration of such a design is that we distribute the semantic matching process to remote devices, in a parallel-computing form, which largely relieves the data processing burden for the source platform. Traditional semantic technologies can be applied to implement this modular such as Web Service Description Language–Semantics (WSDL-S). Another advantage for this framework is that it is independent of communication protocols that reduce the difficulty for implementation.

5.5.5 Runtime Migration Demonstration

Supporting multiplatform interaction migration means integrating platform information and creating appropriate interaction under the correct circumstances. Our migration service resides on each device side, composing a peer-to-peer structure that avoids static registration and supports dynamic device awareness. Such a design, though heavy for small devices [35], indicates its potentiality—especially with the development of the data processing capacity of embedded devices. To affirm practicality, our framework is designed to meet three main requirements: dynamic device awareness, user preference, and usability criterion. To meet the usability criterion and relieve the burden of data processing on the client side, we follow two principles in our designation:

1. *Less data storage on each device.* Each device only stores its own platform information. Device awareness is accomplished by using negotiation

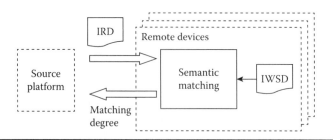

Figure 5.15 Semantic matching.

among devices. Based on this mechanism, device descriptions are not fetched by other devices, thus saving storage.

2. *Less data processing on each device.* Task model tools such as TERESA may surpass the process ability for current small devices. Therefore, we recommend semantic analysis among Web service descriptions for device selection. Our distributed semantic matching process also follows this principle to ease the pressure on client side.

With these design choices, our migration process consists of the following main steps (the sequence of output interaction migration is demonstrated in Figure 5.16):

1. *Request for interaction migration.* Application requests for interaction migration through APIs provided by MSE: The MSE receives interaction elements from the application and combines with user preference to formalize interaction constraints. Requests can also specify the moment for interaction migration (if not specified, interaction will immediately validate). User preference can be obtained by manual input or by physical environment parameters such as position relationship between user and devices detected by GPS, wireless network, or static configurations. After extracting an IRD from requests, the MSE sends an interaction message to devices in the vicinity, asking for device information.

2. *Interaction devices selection.* Device information is represented by Web description language (such as WSDL-S): The IDM is in charge of the semantics of

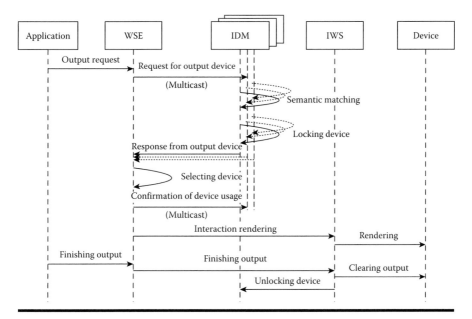

Figure 5.16   The migration process.

these Web services. After receiving an IRD from a source platform, all IDMs calculate semantic matching degrees according to the IRD, lock themselves, and return the results. The MSE collects these semantic matching degrees from the devices and selects the highest matches from among them. The MSE then multicasts the matching results to these devices to establish an interaction connection. The MSE also notifies other devices not chosen to release their locks.

3. *Interaction rendering.* When the MSE determines the interaction devices within the networks, remote devices represented by matched Web services will receive messages from the source platform. Interaction contents are rendered for remote devices in the messages to validate by calling device drives. The IRD will also be transmitted to remote devices through the network. For GUI, remote devices can deploy auto-generation tools to create a GUI based on the interaction elements given. Moreover, call-back instances will also be transmitted to remote devices for interaction feedback.

4. *Interaction termination.* In our framework design, the display moment for interaction elements is determined by program logic. For the interaction output process, the application invokes the interaction termination API provided by the MSE to instruct remote devices to clear output. For the interaction input process, the interaction finishes once input results are returned from remote devices to the source platform. Resource locks on remote devices are released after the interaction's termination.

For input interaction migration, the whole process is similar except that input results will return to the application at the end of the process. We can also divide the migration process according to physical positions. The application and MSE objects to the left side of the sequence graph reside on the source platform, which requests the migration. The IDMs, IWS, and device objects to the right side of the sequence graph reside on target platforms (or remote devices), which receive requests and provide interaction.

5.6 Summary

We proposed a service-oriented framework for HCI migration and provide a context-driven method for service selection within a multimodal environment. Our method extracts user preferences from users' interaction history to describe user-expecting devices when interacting. Moreover, two procedures—local service matching and global combination selection—are proposed to discover the best combination of interaction services. By integrating user preferences, service, and environment contexts, our service matching algorithm is intelligent enough to handle context changes and to provide a better user experience. We also presented a Web service–based HCI migration framework, which avoids centralized servers, and is thus able to support interactions with dynamically emerging and disappearing devices.

Future work will be undertaken to improve continuity of migrated interaction and conflict resolution for busy devices. We are also interested in applying a more flexible service description format to support more accurate semantics and user-customizing service types.

5.7 Appendices

5.7.1 Request Format in XML

```xml
<?xml version="1.0"?>
<xsd:schema xmlns:xsd="http://www.w3.org/2001/XMLSchema">
<xsd:element name=" Request " />
  <xsd:complexType>
    <xsd:sequence>
      <xsd:element name="App" type="xsd:string"/>
      <xsd:element name="UserID" type="xsd:string"/>
      <xsd:element ref="Video Output Service " />
      <xsd:element ref="Audio Output Service " />
      <xsd:element ref="Video Input Service " />
      <xsd:element ref="Audio Input Service " />
    </xsd:sequence>
  </xsd:complexType>
</xsd:element>
<xsd:element name=" Video Output Service" />
  <xsd:complexType>
    <xsd:sequence>
      <xsd:element ref=" Properties" />
      <xsd:element ref=" Preferences " />
    </xsd:sequence>
  </xsd:complexType>
</xsd:element>
<xsd:element name=" Properties " />
  <xsd:complexType>
    <xsd:sequence>
      <xsd:element name="Format" type="xsd:string"/>
      <xsd:element name="Resolution" type="xsd:string"/>
      <xsd:element name="Bandwidth" type="xsd:string"/>
      <xsd:element name="Others" type="xsd:string"/>
    </xsd:sequence>
  </xsd:complexType>
</xsd:element>
<xsd:element name=" Preferences " />
```

```
<xsd:complexType>
  <xsd:sequence>
    <xsd:element name="Device Info" type="xsd:string"/>
    <xsd:element name="Others" type="xsd:string"/>
  </xsd:sequence>
</xsd:complexType>
</xsd:element>
</xsd:schema>
```

5.7.2 Context Format in XML

```
<?xml version="1.0"?>
<xsd:schema xmlns:xsd="http://www.w3.org/2001/XMLSchema">
<xsd:element name=" Context " />
  <xsd:complexType>
    <xsd:sequence>
      <xsd:element ref="Environment Context"/>
      <xsd:element ref="User Context" />
    </xsd:sequence>
  </xsd:complexType>
</xsd:element>
<xsd:element name=" Environment Context" />
  <xsd:complexType>
    <xsd:sequence>
      <xsd:element name="Time " type="xsd:string"/>
      <xsd:element name="Temperature" type="xsd:string"/>
      <xsd:element name="Brightness" type="xsd:string"/>
      <xsd:element name="Others" type="xsd:string"/>
    </xsd:sequence>
  </xsd:complexType>
</xsd:element>
<xsd:element name=" User Context" />
  <xsd:complexType>
    <xsd:sequence>
      <xsd:attribute ref="User Identification"/>
      <xsd:element name="Position" type="xsd:string"/>
      <xsd:element name="Orientation" type="xsd:string"/>
      <xsd:element name="Others" type="xsd:string"/>
    </xsd:sequence>
  </xsd:complexType>
</xsd:element>
</xsd:schema>
```

5.7.3 Device Profile in XML

```
<?xml version="1.0"?>
<xsd:schema xmlns:xsd="http://www.w3.org/2001/XMLSchema">
<xsd:element name=" Profile" />
  <xsd:complexType>
```

```
    <xsd:sequence>
      <xsd:element ref="Video Output Service" />
      <xsd:element ref="Audio Output Service" />
      <xsd:element ref="Keyboard Input Service" />
    </xsd:sequence>
  </xsd:complexType>
</xsd:element>
<xsd:element name="Video Output Service" />
  <xsd:complexType>
    <xsd:sequence>
      <xsd:element ref="Properties" />
      <xsd:element ref="Context" />
    </xsd:sequence>
  </xsd:complexType>
</xsd:element>
<xsd:element name=" Properties " />
  <xsd:complexType>
    <xsd:sequence>
      <xsd:element name="Others" type="xsd:string"/>
    </xsd:sequence>
  </xsd:complexType>
</xsd:element>
<xsd:element name=" Context " />
  <xsd:complexType>
    <xsd:sequence>
      <xsd:element name="Position" type="xsd:string"/>
      <xsd:element name="Orientation" type="xsd:string"/>
      <xsd:element name="Others" type="xsd:string"/>
    </xsd:sequence>
  </xsd:complexType>
</xsd:element>
</xsd:schema>
```

Further Readings

Human–Computer Interaction

J. A. Jacko and A. Sears, Eds., *The Human-computer Interaction Handbook: Fundamentals, Evolving Technologies and Emerging Applications.* Hillsdale, NJ: L. Erlbaum Associates Inc., 2003.

This book provides HCI theories, principles, advances, and case studies, and it captures the current and emerging subdisciplines within HCIs related to research, development, and practice.

Multimodal Interaction
http://www.w3.org/2002/mmi/

This is W3C multimodal interaction workgroup homepage providing information such as the definition of multimodal interaction and multimodal architecture and the specifications of multimodal interaction (e.g., EMMA, InkML, and EmotionML).

ConcurTaskTrees
http://giove.cnuce.cnr.it/tools/CTTE/CTT_publications/index.html
ConcurTaskTrees is a notation for task model specifications used to design interactive applications. This site provides a variety of information about the ConcurTaskTrees features, tools to develop task models, and related publications.

Web Service
https://ws.apache.org/
This page contains information about several Apache projects related to Web services such
as Axiom, SOAP, XML-RPC, and WSS4J.

WSDL
http://www.w3.org/TR/wsdl20/
WSDL is an XML language for describing Web services. This page describes the Web
Services Description Language Version 2.0 (WSDL 2.0).

Semantic Matching
http://semanticmatching.org/
This site provides information about concepts, technologies, tools, and even evaluation
datasets of semantic matching.

References

1. M. Weiser, The computer for the 21st century, *Scientific American,* vol. 265, no. 3, pp. 94–104, 1991.
2. R. Bandelloni, F. Paternò, Flexible interface migration, in *Proceedings of ACM IUI,* 2004.
3. A. Sarmento, *Issues of Human Computer Interaction*, IRM Press, 2004.
4. F. Paternò, *Model-Based Design and Evaluation of Interactive Applications*, Springer-Verlag, Berlin, Germany, 1999.
5. K. Luyten, J. Van den Bergh, C. Vandervelpen, K. Coninx, Designing distributed user interfaces for ambient intelligent environments using models and simulations, *Computers and Graphics*, vol. 30, no. 5, pp. 702–713, 2006.
6. Y. Shen, M. Wang, M. Guo, Towards a web service based HCI migration frame-work, in *Proceedings of the 6th International Conference on Embedded and Multimedia Computing (EMC-11)*, 2011.
7. F. Paternò, C. Santoro, A. Scorcia, A migration platform based on web services for migratory web applications, *Journal of Web Engineering*, vol. 7, no. 3, pp. 220–228, 2008.
8. R. Chinnici, J.-J. Moreau, A. Ryman, S. Weerawarana, Web services description language (WSDL) version 2.0 part 1: Core language, *World Wide Web Consortium, Recommendation REC-wsdl20-20070626*, June 2007.
9. R. Khalaf, N. Mukhi, S. Weerawarana, In Proc. of the WWW Conference (Alternate Paper Tracks), Budapest, Hungary, May 20–24, 2003.
10. S.A. McIlraith, T.C. Son, H. Zeng, Semantic web services, *IEEE Intelligent Systems*, vol. 16, no. 2, pp. 46–53, 2001.

11. D.L. Martin, M. Paolucci, S.A. McIlraith, M.H. Burstein, D.V. McDermott, D.L. McGuinness, B. Parsia, et al. Bringing semantics to web services: The owl-s approach, in *Proceedings of SWSWPC*, pp. 26–42, 2004.
12. M. Klusch, B. Fries, K.P. Sycara, Automated semantic web service discovery with owls-mx, in *Proceedings of AAMAS*, pp. 915–922, 2006.
13. A. Bandara, T.R. Payne, D. De Roure, T. Lewis, A semantic framework for priority-based service matching in pervasive environments, in *Proceedings of OTM Workshops*, pp. 783–793, 2007.
14. Z. Wu, Q. Wu, H. Cheng, G. Pan, M. Zhao, J. Sun, Scudware: A semantic and adaptive middleware platform for smart vehicle space, *IEEE Transactions on Intelligent Transportation Systems*, vol. 8, no. 1, pp. 121–132, 2007.
15. S.B. Mokhtar, A. Kaul, N. Georgantas, V. Issarny, Efficient semantic service discovery in pervasive computing environments, in *Proceedings of Middleware*, pp. 240–259, 2006.
16. D. Chakraborty, A. Joshi, Y. Yesha, T.W. Finin, Toward distributed service discovery in pervasive computing environments, *IEEE Transactions on Mobile Computing*, vol. 5, no. 2, pp. 97–112, 2006.
17. A.K. Dey, *Providing Architectural Support for Building Context-Aware Applications*, PhD thesis, Georgia Institute of Technology, College of Computing, Atlanta, GA, 2000.
18. B. Medjahed, Y. Atif, Context-based matching for web service composition, *Distributed and Parallel Databases*, vol. 21, no. 1, pp. 5–37, 2007.
19. C. Lee, A. Helal, Context attributes: An approach to enable context-awareness for service discovery, in *Proceedings of SAINT*, pp. 22–30, 2003.
20. P. TalebiFard, V.C.M. Leung, Context-aware mobility management in heterogeneous network environments, *Journal of Wireless Mobile Networks, Ubiquitous Computing, and Dependable Applications*, vol. 2, no. 2, pp. 19–32, 2011.
21. F. Morvan, A. Hameurlain, A mobile relational algebra, *Mobile Information Systems*, vol. 7, no. 1, pp. 1–20, 2011.
22. S. Gao, J. Krogstie, K. Siau, Developing an instrument to measure the adoption of mobile services, *Mobile Information Systems*, vol. 7, no. 1, pp. 45–67, 2011.
23. L. Zeng, B. Benatallah, A.H.H. Ngu, M. Dumas, J. Kalagnanam, H. Chang, Qos-aware middleware for web services composition, *IEEE Transactions on Software Engineering*, vol. 30, no. 5, pp. 311–327, 2004.
24. R. Aggarwal, K. Verma, J. Miller, W. Milnor, Constraint driven web service composition in meteors, in *Proceedings of Services Computing*, pp. 23–30, 2004.
25. J. Xiao, R. Boutaba, QoS-aware service composition and adaptation in autonomic communication, *IEEE Journal on Selected Areas in Communications*, vol. 23, no. 12, pp. 2344–2360, 2005.
26. H. Yu, S. Reiff-Marganiec, Automated context-aware service selection for collaborative systems, in *Proceedings of CAiSE*, 2009.
27. K. Luyten, C. Vandervelpen, J. Van den Bergh, K. Coninx, Context-sensitive user interfaces for ambient environments: Design, development and deployment, in *Proceedings of Mobile Computing and Ambient Intelligence: The Challenge of Multimedia*, 2005.
28. K. Luyten, C. Vandervelpen, K. Coninx, Task modeling for ambient intelligent environments: Design support for situated task executions, in *Proceedings of TAMODIA*, 2005.

29. R. Han, V. Perret, M. Naghshineh, Websplitter: A unified XML framework for multi-device collaborative web browsing, in *Proceedings of CSCW*, pp. 221–230, 2000.

30. R. Bandelloni, F. Paternò, Platform awareness in dynamic web user interfaces migration, in *Human-Computer Interaction with Mobile Devices and Services*, pp. 440–445, Springer, Berlin, 2003.

31. F. Paternò, G. Mori, R. Galiberti, CTTE: An environment for analysis and development of task models of cooperative applications, in *Proceedings of ACM CHI*, pp. 21–22, 2001.

32. G. Mori, F. Paternò, C. Santoro, CTTE: Support for developing and analyzing task models for interactive system design, *IEEE Transactions on Software Engineering*, vol. 28, no. 8, pp. 797–813, 2002.

33. A.R. Puerta, J. Eisenstein, Towards a general computational framework for model-based interface development systems, in *Proceedings on IUI*, pp. 171–178, 1999.

34. G. Mori, F. Paternò, C. Santoro, Tool support for designing nomadic applications, in *Proceedings of IUI*, pp. 141–148, 2003.

35. R. Bandelloni, F. Paternò, Flexible interface migration, in *Proceedings of IUI*, pp. 148–155, 2004.

36. H.H. Hsu, C.C. Chen, RFID-based human behavior modeling and anomaly detection for elderly care, *Mobile Information Systems*, vol. 6, no. 4, pp. 341–354, 2010.

37. A. Jaimes, N. Sebe, Multimodal human-computer interaction: A survey, in *Proceedings of ICCV Workshop on HCI*, 2005.

38. E. Hjelmås, B.K. Low, Face detection: A survey, *Computer Vision and Image Understanding*, vol. 83, no. 3, pp. 236–274, 2001.

39. B. Fasel, J. Luettin, Automatic facial expression analysis: A survey, *Pattern Recognition*, vol. 36, no. 1, pp. 259–275, 2003.

40. J.K. Aggarwal, Q. Cai, Human motion analysis: A review, *Computer Vision and Image Understanding*, vol. 73, no. 3, pp. 428–440, 1999.

41. A.T. Duchowski, A breadth-first survey of eye-tracking applications, *Behavior Research Methods, Instruments, & Computers*, vol. 34, no. 4, pp. 455–470, 2002.

42. Y.H. Han, C.M. Kim, J.M. Gil, A greedy algorithm for target coverage scheduling in directional sensor networks, *Journal of Wireless Mobile Networks, Ubiquitous Computing, and Dependable Applications*, vol. 1, no. 2/3, pp. 96–106, 2010.

43. I. You, T. Hara, Mobile and wireless networks, *Mobile Information Systems*, vol. 6, no. 1, pp. 1–3, 2010.

44. M. Wang, X. Tang, Y. Shen, M. Guo, A method of context-driven HCI service selection in multimodal interaction environments, in *Proceedings of the Second International Symposium on Frontiers in Ubiquitous Computing, Networking and Applications*, 2011.

45. P. Pyla, M. Tungare, J. Holman, M. Pérez-Quiñones, Continuous user interfaces for seamless task migration, in *Proceedings of Human-Computer Interaction*, 2009.

Chapter 6

Pervasive Mobile Transactions

Pervasive computing is a user-centric distributed computing paradigm that allows users to transparently access their preferred services even while moving around. Furthermore, many pervasive applications (e.g., a fund transfer between two different banking systems) must be executed reliably. Pervasive transaction processing can be used to guarantee such transparent and reliable services for users.

In this chapter, Section 6.1 introduces the concept of pervasive transactions. Section 6.2 presents a pervasive transaction processing framework, followed by a context-aware pervasive transaction model in Section 6.3. Then we propose a context-adaptive dynamic transaction management algorithm in Section 6.4. Finally, Section 6.5 models and verifies the accuracy of the proposed transaction management algorithm using Petri nets.

6.1 Introduction to Pervasive Transactions

Owing to the high mobility of users, pervasive systems run in an extremely dynamic and heterogeneous environment, where various networked devices and communication protocols coexist, and network topology and bandwidth are dynamically mutative and frequent [1–5]. Therefore, open pervasive environments are prone to various failures from networks, devices, applications, and basic services of pervasive systems [3]. To facilitate use as much as possible, pervasive systems have to intelligently handle failures and hide complex recovery processes from users, providing transparent communication and computing services [1,3]. As a result, transaction management technology, which has been widely used to guarantee system consistency for a variety of distributed computing paradigms, is an important enabling technology for making user-centric pervasive computing a reality.

Context adaptation is an important characteristic that distinguishes pervasive systems from traditional distributed environments. Specifically, pervasive environments exhibit the following features:

- *High mobility.* Pervasive systems provide mobile users with transparent online services. With the movement of a node, neighbors change continuously. As a result, a context-aware communication mechanism is an important feature of pervasive transaction processing.
- *Limited resources.* Networked mobile devices have limited energy, computing capacity, and storage resources. Therefore, energy-efficient communication and computation are important considerations [6,7].
- *Uncertain fixed infrastructure.* Traditional mobile transaction models rely on wired fixed networks. In pervasive environments, however, networks cover almost every area, and computing and service extend from fixed networks to various wireless networks. A fixed infrastructure, such as a base station, is not a requisite any more.
- *Heterogeneous and changing environments.* Networks, data, and devices are different from each other in a pervasive environment. Various wireless networks with different communication protocols (e.g., Bluetooth, Wi-Fi) coexist with wired networks, while at the same time the network connection and bandwidth are extremely unstable.

These features require that pervasive transactions have to be aware of time- and space-changing context and therefore must adjust execution behaviors dynamically, as is shown in Figure 6.1. For example, let each hospital in a city scatter a few mobile medical assistants (MMA) equipped with mobile devices that keep on moving around the city to provide medical treatment assistance to potential patients. A traveler (Alice) is driving a car with a mobile device (MD), suddenly suffers from an acute illness, and urgently needs to find a hospital. First, Alice discovers and then requests the nearest MMA (T_1). The MMA checks its lightweight mobile database and reserves a sickbed that meets

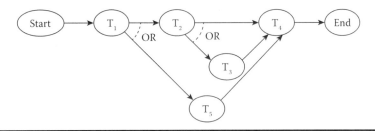

Figure 6.1   A scenario of pervasive transactions.

Alice's requirements (T_2). If the MMA does not hold the corresponding data (e.g., a currently available sickbed) in the local database, it downloads the data from the central database on a fixed network or transfers the reservation request to the central database server (T_3). After a sickbed is successfully reserved, Alice queries the public traffic service to find the best route to that hospital (T_4). If Alice cannot find out any MMA within a given distance, she queries the Health Service Central Hospital for a sickbed reservation (T_5); although that hospital is far away, it can provide a variety of medical services.

Although there have been many transaction models proposed for traditional mobile systems [8–11], they focused on variable bandwidth, network disconnection, replication and synchronization, and hand off of mobile hosts (MHs). Clustering [12] (also called weak–strict) groups semantics-related databases within a cluster. Data are kept in two versions: a weak consistency (local consistency) version for weak transactions and a strict consistency (global consistency) version for strict transactions. Strict and weak transactions are executed when a MH is either strongly connected or weakly connected with fixed networks, respectively. Similar to clustering, two-tier replication [13] (also called base–tentative) maintains a master copy for base transactions and multiple replicated copies for tentative transactions. High Commit Mobile (HiCoMo) transactions [14] keep two kinds of tables: base tables for base transactions and aggregate tables for HiCoMo transactions. An Isolation Only Transaction (IOT) [15] also includes two kinds of transactions—for MH states that are connected and disconnected.

In mobile transaction coordination, the Moflex model [16] submits subtransactions to mobile transaction managers running in base stations, which then submits the subtransactions to databases. Each Moflex transaction is composed of a set of dependency relationships, handoff rules, and expected final states. Kangaroo transaction [17] deploys a data access agent (DAA) in each base station for mobile transaction management. Promotion [18] is a nested, long-lived transaction model where top transactions are executed on fixed hosts and subtransactions on MHs controlled by the compact. In pre-serialization [19], each base station runs a global transaction coordinator responsible for transaction coordination as well as disconnect and mobility management. TCOT [20] is a one-phase commit protocol for a transaction termination decision (e.g., commit, abort, etc.) in message-oriented systems. This protocol uses a time-out mechanism to reduce the impact of a slow and unreliable wireless link, where the time-out not only enforces the termination condition but also the entire execution duration as well. Lim and Hurson [21] proposed a hierarchical concurrency control algorithm—v-lock, which uses global locking tables to serialize global transactions and to detect and remove global deadlocks. The global locking tables are created with semantic information contained within the hierarchy. A data replication scheme, which caches queries and the associated data at the mobile unit as a complete object to alleviate the limited bandwidth and local autonomy restrictions, was also used in this research.

These existing mobile transaction models cannot be directly applied to mobile pervasive systems because of the following:

1. *Insufficient support for the mobility.* Traditional mobile transaction models are built on a client–server–proxy architecture [22], where fixed hosts are data sources and transaction servers. Therefore, these models only address the mobility of clients without considering the mobility of data servers. In pervasive environments, however, mobile devices interact with each other in a peer-to-peer way, and both the client and the server will be moving during transaction processing.
2. *Little consideration of context awareness.* Pervasive transaction management has to adapt to transaction context. Existing proposals have not solved such an issue.

6.2 Mobile Transaction Framework

Existing client–proxy–server-based mobile transaction models need base stations for communication and transaction management and fixed data servers for transaction execution. Therefore, they only address the mobility of clients. In pervasive environments, however, users and devices keep moving, so there are no fixed base stations. Instead, mobile devices can work as both clients and servers in a peer-to-peer way. As a result, traditional mobile transaction modes will face the following *unavailable transaction service* problem.

6.2.1 Unavailable Transaction Service

In traditional client–proxy–server-based mobile transaction systems, base stations in the fixed network are at the center and work as gateways among mobile clients as well as fixed servers and mobile transaction managers. An MH can communicate with a base station only when it is located within the corresponding cell. Specifically, the MH disconnects with any database server if the MH moves out of the range of any base station, causing the *unavailable transaction service* as illustrated in Figure 6.2.

The unavailable transaction service problem seriously restricts online pervasive services. But its online service is one of the most attractive scenes distinguishing pervasive computing from traditional mobile computing paradigm. So, a new transaction processing framework should be designed for pervasive transactions that enable MHs to communicate with each other anytime, anywhere.

6.2.2 Pervasive Transaction Processing Framework

To make the provision of pervasive transaction service anytime and anywhere a reality, we propose a new pervasive transaction processing framework, as shown in Figure 6.3. In this framework, neighboring mobile devices that use the same

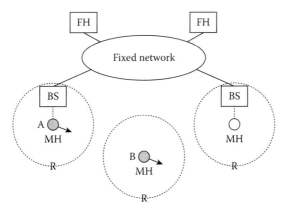

Figure 6.2   Unavailable transaction service. MH: mobile host; BS: base station; FH: fixed host; R: ratio range of a mobile host.

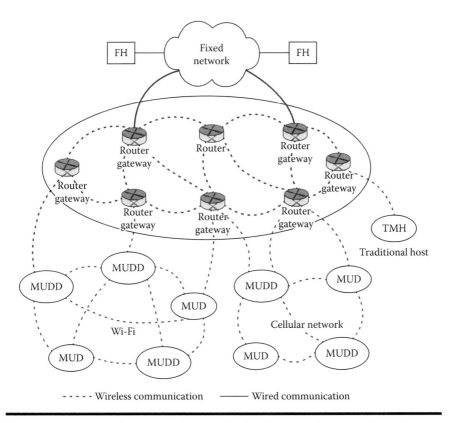

Figure 6.3   A pervasive transaction processing framework.

wireless communication protocols automatically connect with each other. In such a framework, there are two kinds of nodes: a mobile ubiquitous device without any database (MUD), such as a smart mobile phone, and a mobile ubiquitous device with database(s) (MUDD), such as a powerful laptop. Accordingly, MUDDs work not only as clients initiating mobile transactions but also as servers executing transactions while MUDs act only as clients.

The pervasive transaction processing framework is composed of three layers of networks: *fixed network, wireless routing backbone network,* and *wireless mesh subnetwork.* Each node forwards data for other neighboring nodes so that any two nodes can connect by multiple hops. In Figure 6.3, dashed and solid lines denote wireless and wired communications, respectively; thicker lines mean higher wireless bandwidths; "FH" denotes a fixed host connected with a fixed network; "MUD" and "MUDD" mean mobile ubiquitous devices without any database and with database(s), respectively; and "TMH" indicates a traditional MH without mesh functionalities. MUDs and MUDDs with identical communication protocols are interconnected into low-level wireless mesh networks automatically and dynamically.

A pervasive transaction can be initiated by any mobile device, with the following different execution models:

■ A pervasive transaction is entirely executed by MUDD(s), which occurs more often than the next two models.
■ A pervasive transaction is distributed between MUDD(s) and FHs.
■ A pervasive transaction is entirely executed by FHs.

6.3 Context-Aware Pervasive Transaction Model

Pervasive transaction processing must be aware of changing context. In this section, we will present a context model and then a context-aware transaction model because they are prerequisite to adaptive transaction management.

6.3.1 Context Model for Pervasive Transaction Processing

Pervasive transaction context covers all the information relevant to individuals, networks, and devices within the activity space, specifically including the following dimensions:

■ *Person*: profile, preferences, and requirements of users
■ *Wireless network*: connectivity and performance of networks
■ *Mobile device*: computing and storage capacity of mobile devices
■ *Location*: longitude and latitude (or relative position) of mobile devices.

Table 6.1   Context Details of Pervasive Transactions

Entity	Attribute	Value
Person	Name Sex Age Requirement Preference	User name Male or female Value of age Requirement description Behavior preference
Wireless networks	Connectivity Bandwidth Delay Lost_ratio Cost Stability	Connected, disconnected High, medium, low High, medium, low High, medium, low Expensive, cheap, free Good, medium, bad
Mobile devices	Available_battery Available_data Computing_capacity Available_memory Available_cache Security	Full, half, low Available, unavailable High, medium, low Full, half, low Full, half, low Secure, open
Location	Location	Pairs of longitude and latitude

Table 6.1 lists context details of pervasive transactions; this is where some attributes may be described more accurately. For example, we can represent computing capacity by the main frequency of a mobile device and available memory by the size of available storage.

A context model is a formal representation of physical space, information space, and human activities. We abstract transaction context using *entities*, *attributes*, and *relationships* among entities and model the context using an entity–relationship diagram. The graphic context model for pervasive transactions is shown in Figure 6.4, where we only illustrate attributes of the entity, *person*. Attributes of other entities can be easily complemented in the same way as in Table 6.1. Some attributes (e.g., preference of a user) cannot be directly abstracted from entity profiles, and data mining–type technologies have to be used for this purpose. Entities are described through pairings of attributes and values.

6.3.2 Context-Aware Pervasive Transaction Model

A context-aware transaction model should be able to adjust behaviors dynamically in terms of changing transaction context. A context-aware pervasive transaction can be formally defined as follows.

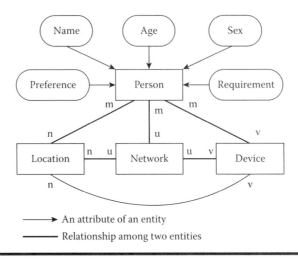

Figure 6.4 Context model for pervasive transactions.

Definition 6.1

A pervasive transaction is a 7-tuple (T, CT, ECA, ECA$_i$, D, TS, FS), where:

- T = {T$_i$ | 1 ≤ i ≤ n} is the set of all subtransactions in a pervasive transaction.
- CT = {CT$_i$ | 1 ≤ i ≤ n} is the set of compensating transactions for all the subtransactions.
- ECA = <ECA$_i$> is the list of ECA rules for all the subtransactions.
- ECA$_i$ = <E$_i$, C$_i$, A$_i$> is the list of 3-tuple with event, context, and action for the subtransaction T$_i$.
- D is the set of interdependencies that specify relationships between T$_i$ and T$_j$ (T$_i$, T$_j$ ∈ T).
- TS = { S$_i$ | 1 ≤ i ≤ n} is the set of states of all subtransactions.
- FS is the set of acceptable final states.

The formula ECA = <ECA$_i$ | 1 ≤ i ≤ n > is at the center of the context-aware pervasive transaction mode. ECA *consists of* a list of ECA rule descriptors, where ECA$_i$ is a rule descriptor for a subtransaction T$_i$, and ECA$_i$ has a higher priority than ECA$_{i+1}$. In particular, each ECA$_i$ = <E$_i$, C$_i$, A$_i$> is also a list of 3-tuple (event, context, and corresponding action), describing multiple context-based execution policies for a subtransaction T$_i$, where

- E$_i$ = {E$_{ij}$} is the set of events that occur during the execution of T$_i$.
- C$_i$ = {C$_{ik}$} is the set of context associated with a subtransaction T$_i$. We take C$_i$ as the conditions that executing T$_i$ has to meet. C$_i$ covers four dimensions: WN, MD, location, and person, described in Table 6.1.
- A$_i$ = {A$_{il}$} is the set of corresponding actions.

Table 6.2   Transaction State Description

Symbol	State	Description
I	Initiation	T_i does not start to execute
E	Executing	T_i is executing and has not committed
S	Committed	T_i has successfully committed
F	Failed	T_i has failed to commit and is rolled back to previous state
C	Compensating	Compensating transaction C_i has been executing

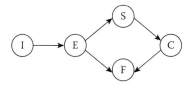

Figure 6.5   State conversion diagram of transactions. I, E, S, F, and C represent five states, Initiation, Executing, Committed, Failed, and Compensating, respectively.

D is the set of intradependencies between T_i and T_j (T_i, $T_j \in T$). We define two kinds of intradependencies: *success dependency* ($\prec_s$) and *failure dependency* ($\prec_f$). $T_i \prec_s T_j$ means that T_j cannot be executed until T_i successfully commits. $T_i \prec_f T_j$ denotes that T_j can start to execute only if T_i fails to commit or is compensated in the case that T_i has committed.

TS = {S_1, S_2, …, S_n} is the set of states of all subtransactions in a pervasive transaction. S_i, the state of subtransaction T_i, is one of five possible states: I, E, F, S, and C (see Table 6.2). Each subtransaction starts with state "I" and ends with state "S" (T_i is successfully committed) or "F" (T_i is aborted). Figure 6.5 illustrates the state conversion of transactions.

FS is the set of acceptable final states. A pervasive transaction may have multiple proper final states acceptable to a user, probably with different priorities. For example, Alice first discovers the quickest qualified devices for the sickbed reservation and, after a failure, she will contact the Health Service Central Hospital to reserve a sickbed.

6.3.3 A Case of Pervasive Transactions

We model the medical treatment reservation scenario mentioned in Section 6.1 as a mobile pervasive transaction (T, CT, ECA, D, TS, FS). MD is a mobile device used by Alice; MMA is a mobile medical assistant that is equipped with mobile device(s) and keeps on moving around. According to Section 6.1, we have:

■ T = {T_1, T_2, T_3, T_4, T_5}. The global transaction T consists of five subtransactions. For the meaning of each subtransaction, please refer to the scenario in Section 6.1.

- $CT = \{\varnothing, CT_2, CT_3, \varnothing, CT_5\}$. CT_2, CT_3, and CT_5 are compensating transactions of T_2, T_3, and T_5, respectively. T_1 and T_4 do not need to be compensated.
- $D = \{T_1 <_s T_2, T_2 <_s T_4, T_2 <_f T_3, T_3 <_s T_4, T_1 <_f T_5, T_5 <_s T_4\}$. The intradependency D specifies that T may be executed in one of three sequences: $\sigma_1 = \{T_1, T_2, T_4\}$, $\sigma_2 = \{T_1, T_3, T_4\}$, or $\sigma_3 = \{T_5, T_4\}$.
- $TS = \{S_1, S_2, \ldots, S_5\}$. S_i is the state of subtransaction T_i. The transaction initiator MD updates TS once a subtransaction is executed.
- $FS = \{(S, S, -, S, -), (S, F, S, S, -), (F, -, -, S, S)\}$, where "S" and "F" stands for a successful commit and an abort of the corresponding subtransaction, respectively, while "-" means that the execution state of the subtransaction does not affect the decision.

ECA describes how to adapt the transaction context and thus is a central element of our pervasive transaction model. In this example, ECA consists of five rule descriptors (each for a subtransaction) such that $ECA = <ECA_1, ECA_2, ECA_3, ECA_4, ECA_5>$, where

- $ECA_1 = <<E_1, C_{11}, A_{11}>, <E_1, C_{12}, A_{12}>>$. $<E_1, C_{11}, A_{11}>$, and $<E_1, C_{12}, A_{12}>$ describe that MD successfully connects with and fails to discover the nearest MMA, where E_1 means Alice tries to discover the nearest MMA; C_{11} means the network link and bandwidth among the MD and qualified MMAs are good enough for communication; and a discovered MMA is able to handle the transaction request; A_{11} means the MD selects the MMA as a transaction participant, on behalf of Alice; C_{12} means the network link or bandwidth between the MD and any MMA is not qualified for communication; and A_{12} refers to when MD connects with the Health Service Central Hospital. Note that in that case, subtransaction T_1 has failed.
- $ECA_2 = <E_2, C_2, A_2>$. The MMA checks its local database to reserve a sickbed meeting Alice's requirements. If the data are locally available, T_2 can successfully commit. Otherwise, T_2 will fail.
- $ECA_3 = <<E_{31}, C_{31}, A_{31}>, <E_{32}, C_{32}, A_{32}>>$. In a case where the MMA cannot find the needed data in its local database, it continues to handle the sickbed reservation. E_{31} means subtransaction T_2 failed; C_{31} means the MMA has enough computing and storage capacity; A_{31} means the MMA downloads the needed data from the central database; E_{32} means the MMA finds that it is not able to store the needed data; C_{32} means the link between the MMA and the central database server is qualified; and A_{32} refers to when the MMA transfers the sickbed reservation request to the central database server.
- $ECA_4 = <E_4, C_4, A_4>$. After finishing the sickbed reservation, MD queries a public traffic information service to discover the best path to that hospital.
- $ECA_5 = <E_5, C_5, A_5>$. MD requests the Health Service Central Hospital reserve a sickbed. We assume the central database server of the Health Service Central Hospital can handle transactional medical treatment requests all the time.

6.4 Dynamic Transaction Management

Due to changing pervasive context, pervasive transaction management has to adaptively adjust execution policies during transaction processing, which will be presented in this section.

6.4.1 Context-Aware Transaction Coordination Mechanism

A pervasive transaction is initiated by a mobile device (MD or MDD) called an *initiator*. Then, in a multiple-hop fashion, the initiator distributes subtransactions to neighboring MDDs or FHs, called *executors*. In wireless networks, data communication consumes more energy than application operations. The transaction management mechanism works via the following principles:

- An *initiator* discovers the *shortest* and most *qualified executors*. By *qualified*, we mean the *executors* have enough computing and storage capacity and bandwidth among the *initiator* and the *executors*. *Shortest* means the executor has the least number of hops away from the initiator. The fewer the number of hops, the less energy is consumed for a unit of data transmission. Therefore, this approach reduces the energy consumption of energy-limited mobile devices while it improves the probability of a successful transaction commit at the same time.
- An *executor* (MDD) actually performs application operations in the subtransaction. In general, a resource-limited MDD has a lightweight mobile database that holds partial data from the central databases running on the fixed host. If accessed data are available locally, the executor performs the subtransaction directly. Otherwise, the executor will download the accessed data from the central database or will transfer the subtransaction to the fixed central database server.

6.4.2 Coordination Algorithm for Pervasive Transactions

Let a global pervasive transaction T consists of n subtransactions such that $T = \{T_1, T_2, \ldots, T_n\}$. The state of T is a set of states of all subtransactions such that $TS = \{S_1, S_2, \ldots, S_n\}$, where S_i (i.e., the state of T_i) is one of five possible states: I, E, F, S, and C. TS is updated once a subtransaction is executed. Based on TS, states of global transactions can be divided into the following three types:

- *Acceptable state.* TS reaches one of states acceptable by a user (i.e., $TS \in FS$).
- *Executing state.* TS has not reached any acceptable state, and part of the subtransactions have not been executed such that $TS \notin FS \wedge \exists S_i = \text{'I'}$.
- *Failed state.* TS has not reached any acceptable state, but all subtransaction have been executed such that $TS \notin FS \wedge \forall S_i \neq \text{'I'}$.

MDDs have the ability to handle pervasive transactions. First, the transaction coordination mechanism discovers dynamically qualified devices such as transaction executors and then sends subtransactions to them. If data to be accessed during a subtransaction is locally available, the executor finishes application operations and performs the subtransaction locally. Otherwise, the executor will decide whether it downloads the data from its central database on a fixed host to a local database or transfers the subtransaction to the fixed host. From the system point of view, our transaction management involves an initiator and a set of mobile and/or fixed executors, where the initiator controls the pervasive transaction processing. Figure 6.6 illustrates the execution flow of the pervasive transaction coordination.

An initiator coordinates a pervasive transaction in the following steps:

1. Find (P_i, T_j): An initiator discovers a qualified neighboring node as a participant P_i to execute a subtransaction T_j. It is context-aware which subtransaction will be executed, based on event–context–action rule descriptors defined in pervasive transaction T.

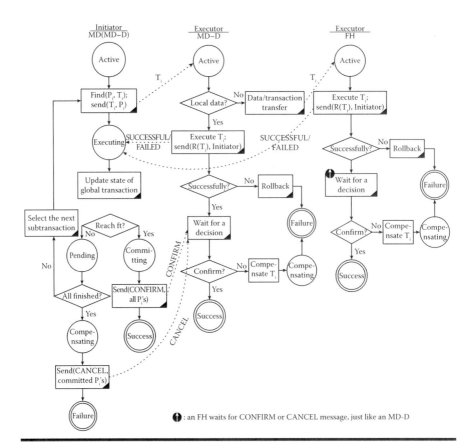

Figure 6.6 Execution process of pervasive transactions.

2. Send (T_i, P_i): The initiator sends T_i to the participant P_i found in the first step.
3. Receive $(R(T_i), P_i)$: The initiator waits for execution result $R(T_i)$ from the P_i. $R(T_i)$ is a SUCCESSFUL or a FAILED message, which means that P_i has successfully committed or failed to commit, respectively.
4. Global decision: Based on the T' current state TS, the initiator makes a global decision:
   - *Global commit.* If T reaches the *acceptable state*, the initiator confirms all committed subtransactions by sending CONFIRM messages to them. At that time, the pervasive transaction T successfully commits.
   - *Global abort.* If T reaches the *failed state*, the initiator undoes the committed subtransactions, sending COMPENSATION messages to participants. In that case, T is aborted.
   - *Continuous execution through selecting the next subtransaction.* If T is still at the *executing state*, the initiator selects the next subtransaction T_j based on the $R(T_i)$ and the intradependency D between T_j and T_i and then goes to step (1).

An executor actually performs the incoming subtransaction T_i under the control of the initiator in the following way:

1. *Data verification.* The executor checks the data that will be accessed during the execution of T_i in its local lightweight database. If there are no such data, the executor goes to step (3).
2. *Subtransaction execution.* If the accessed data are locally available, the executor performs T_i:
   - Executing application operations in T_i
   - Sending a SUCCESSFUL (or FAILED) message to the initiator when T_i is successfully committed (or fails to commit)
   - Executing a compensating transaction and recording a *Failure* in a log if it receives a CANCEL message
   - Recording a *Success* after it receives a CONFIRM message
3. *Data or subtransaction transfer.* In a case where the executor cannot find the data in its own local database, it needs to download the data from the central database server or transfer T_i to the central database server. The central database server refers to the main database of an organization, which holds all data needed by the organization's activities and is connected with a fixed network. When the central database server receives a subtransaction, it executes the subtransaction using the policy described in step (2).

6.4.3 Participant Discovery

In pervasive environments, the centralized registry mechanism is impracticable for pervasive transactions due to the node mobility. Accordingly, before submitting a subtransaction, an initiator has to dynamically find a qualified (mobile) device that is able to provide the specific services (e.g., sickbed reservation) for the subtransaction as a participant.

MDs are generally battery-powered so that power saving is one of the most important performance metrics. Generally speaking, the fewer the number of hops of message transmission in multiple-hop mesh networks, the lower the energy consumption of mobile devices. To reduce total energy consumption, the transaction initiator finds and selects participants based on two principles. The first one is to find a mobile device within the shortest distance. The second is to select the mobile device with the most remaining energy. We use the number of hops between the initiator and a participant to measure the distance. The initiator discovers a qualified participant for a subtransaction T_i through the following mechanism:

1. *Participant query.* An initiator broadcasts a query message *REQ* within the format shown in Figure 6.7a. "SD" is a service description, and only mobile devices that can provide the specified services respond to the REQ message; h is the number of hops between the source node S and a destination node D.
2. *Forwarding and response.* Upon receiving an REQ message, any mobile device
   ■ Appends itself to the *path* field and increases *h* field by 1.
   ■ Constructs and returns a response message RES along with the current device states, shown in Figure 6.7b, only if it can provide services specified in the SD field.
   ■ Otherwise, it further broadcasts the updated RES message if h is less than H_{max}. However, if *h* is equal to H_{max}, it drops the message. Note that any device ignores the messages received previously.
3. *Participant selection.* The initiator collects response messages within a given period of time, and then
   ■ Extracts the number of hops from the h field of each RES message, marking h(k).
   ■ Calculates the minimal number of hops $H_{min} = \underset{\text{for all RES}}{\text{Min}}\left(h(k)\right)$.

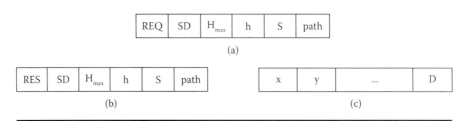

Figure 6.7 Query and response messages for participant discovery: (a) query message, (b) response message and (c) path in the response message.

- Selects a mobile device with H_{min} hops and qualified context (e.g., enough network bandwidth) as a participant. When many devices have the same H_{min} hops, if there is a fixed host with H_{min} hops, the fixed host will be selected as a participant; otherwise, the mobile device with the most remaining energy will be selected as a participant.

4. *Default participant.* In this case, the initiator has failed to find any qualified device within H_{max} hops. We assume that there be well-known fixed hosts to provide public services in a city. If there is no response, which means any mobile device within the range of H_{max} hops cannot execute T_i, the initiator selects a well-known fixed host eligible for execution of T_i.

6.5 Formal Transaction Verification

In this section, we model the aforementioned coordination algorithm through Petri nets and then validate the algorithm's correctness using the Petri nets' reachable tree analysis technology.

6.5.1 Petri Net with Selective Transition

A Petri net is an abstract and formal tool that can model systems' events, conditions, and the relationships among them. The occurrence of these events may change the state of the system, causing some of the previous conditions to cease holding and other conditions to begin to hold [23]. A Petri net consists of two kinds of nodes: places and transitions, which are connected by directed arcs from a place to a transition ($P \times T$) or from a transaction to a place ($T \times P$). In a Petri net, places and transitions represent conditions and events, respectively. In this chapter, we introduce a *selective transition* concept to express the selective activity. Before the definition of selective transition, we assume a transition t has m output arcs such that $O(t) = \{O(t)^{(i)} \mid O(t)^{(i)}$ is one of the output arcs of t, $1 \le i \le m\}$.

Definition 6.2

A transition t is *selective* if (1) any output arc $O(t)^{(i)} \in O(t)$ associates with a condition $O(t)^{(i)}.cond(i)$ and (2) when a firing occurs, only the output arc $O(t)^{(i)}$ with $O(t)^{(i)}.cond(i) = true$ $(1 \le i \le m)$ is fired.

A selective transition concept extends the modeling ability of Petri nets by introducing conditional firing for output arcs. For an activity graph with a conditional loop, shown in Figure 6.8a for example, we model such an activity sequence in a selective Petri net; see Figure 6.8b. When transition t_2 is fired, only the output arc $O(t)^{(i)}$ with $O(t)^{(i)}.cond(i) = true$ is actually fired.

Based on the selective transition concept, we model the aforementioned coordination algorithm as the graphic Petri net (see Figure 6.9). In particular, transaction

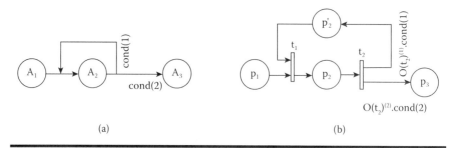

(a)   (b)

Figure 6.8 (a) Conditional activity and (b) selective transition.

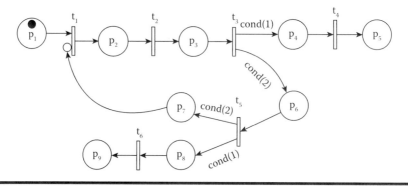

Figure 6.9 Petri net of the coordination algorithm.

t_1 has two exclusively input arcs: $p_1 \times t_1$ and $p_7 \times t_1$, where a white circle means t_1 can be fired if and only if one of the input arcs has a token. Places and transitions are described in Table 6.3. Two selective transitions, t_3 and t_5, denote judging the state of global transaction T and the situation of subtransactions, respectively. More specifically:

■ $O(t_3)^{(1)}$.cond(1): T reaches one of the acceptable states.
■ $O(t_3)^{(2)}$.cond(2): T has not reached any acceptable state.
■ $O(t_5)^{(1)}$.cond(1): All subtransactions have been executed (i.e., $\forall S_i$ = 'S' or 'F').
■ $O(t_5)^{(2)}$.cond(2): Some subtransactions have not executed (i.e., $\exists S_i$ = 'I').

6.5.2 Petri Net–Based Formal Verification

Petri nets can analyze systems' behavioral properties, including reachability, boundedness, liveness, coverability, reversibility, persistence, and so on. For a bounded Petri net, these problems can be solved by the reachable tree [24]. Peterson [23] also pointed out that, in Petri nets, many questions often can be reduced to the reachable problem. In this subsection, we first construct the reachability of the Petri net

Table 6.3 Places and Transitions of Petri Net of the Coordination Algorithm

Elements	Description	Meaning
p_1	Active	A pervasive transaction T is initiated
p_2	Executing	T has been executing
p_3	Waiting	Initiator is receiving the $R(T_i)$
p_4	Committing	Initiator is committing globally
p_5	Success	T is successfully committed
p_6	Pending	T enters the pending state
p_7	Select subtransaction	Initiator selects the next subtransaction based on dependency
p_8	Compensating	Initiator is undoing subtransactions committed previously
p_9	Failure	T has been undone
t_1	Initiator discovers a participant P_i and sends T_i to P_i	
t_2	Initiator updates the state of global transaction T	
t_3	Initiator judges whether T reaches one of the acceptable states	
t_4	T is committed globally	
t_5	Initiator judges whether all subtransactions are executed	
t_6	Subtransactions committed previously are undone by compensating transactions	

and then validate the correctness of the coordination algorithm by the reachable tree analysis technology of Petri nets.

Figure 6.10 illustrates a Petri net reachable tree for the coordination algorithm, which intuitively describes all the states from the initial state M_0 by the movement of tokens. In a reachable tree, a marking M is an assignment of tokens to each place. M is reachable from another marking M' if M' may be transformed to M through a sequence of firings. The set of all markings reachable from an initial marking M_0 in a Petri net (N, M_0) is marked by $R(M_0)$.

The boundedness and liveness of Petri nets are often used as correctness criteria in protocol validation. This chapter discusses these two properties by analyzing a Petri net's reachable tree. In a reachable tree, nodes denote M_0 and its successors; M_0 is the root, and leaf nodes correspond to the final state. A path from the M_0 to a leaf node means an execution sequence. The following theorems collaboratively prove the correctness of the coordination algorithm.

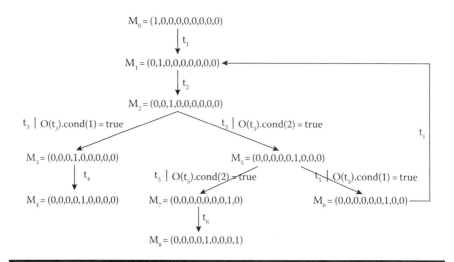

Figure 6.10   Reachable tree of Petri net of the coordination algorithm.

Theorem 6.1

The Petri net of the coordination algorithm is bounded.

Proof: A Petri net (N, M_0) is said to be k-bounded (or simply bounded) if the number of tokens in any one place does not exceed a finite number k for any marking reachable from M_0, that is, $\forall m_i \leq k$ for every place p_i in every marking $M \in R(M_0)$ [24]. In the reachable tree of a Petri net, the number of tokens in each place is never more than 1. Therefore, the Petri net of the coordination algorithm is bounded, and the k is equal to 1.

For a bounded Petri net, its reachable tree contains all possible markings. Let MS be the marking set of the Petri net for the coordination algorithm. We have MS = $R(M_0)$ = {M_0, M_1, M_2, M_3, M_4, M_5, M_6, M_7, M_8}; that is, any marking M in the Petri net of the coordination algorithm is reachable from M_0 such that $M_i \in R(M_0)$ for any $M_i \in$ MS. Therefore, there is no useless or dead state during the execution of a pervasive transaction.

Theorem 6.2

The Petri net of the coordination algorithm is L1-live.

Proof: A transition t is L1-live if t can be fired at least once in some firing sequences. Furthermore, a Petri net (N, M_0) is said to be L1-live if each transition in the net is L1-live [24]. By observing the reachable tree of Petri net of the coordination algorithm, we find that each marking is reachable and each transition can be fired at least once from M_0. Therefore, the Petri net of the coordination algorithm is L1-live.

Theorem 2 indicates that the coordination algorithm is deadlock-free as long as the firing starts with the initial marking M_0. According to Theorem 1 and Theorem 2, we can draw a conclusion: The transaction coordination algorithm is correct.

6.6 Evaluations

This section evaluates the performance of the coordination algorithm through a simulation system, as shown in Figure 6.11, where a *coordinator* and a *participant* execute the transaction coordination mechanism. During the pervasive transaction processing, the *context-awareness* module collects and monitors transaction context for dynamically adjusting transaction execution policies. *Log* service records necessary coordination and state information in order to recover systems from potential failures. A compensating transaction generator (CTG) automatically generates compensating transactions during the execution of pervasive transactions. A *communication unit* sends and receives messages. A local transaction manager (LTM), usually part of a local database, is responsible for ensuring the atomicity, consistency, isolation, and durability (ACID) properties of local subtransactions, while the coordinator manages a global pervasive transaction to achieve one of the acceptable states.

6.6.1 Experiment Environment

In the simulation system, there were 100 mobile devices that simulate MDDs and 10 fixed hosts; 20 mobile devices (i.e., executors) provide sickbed reservation

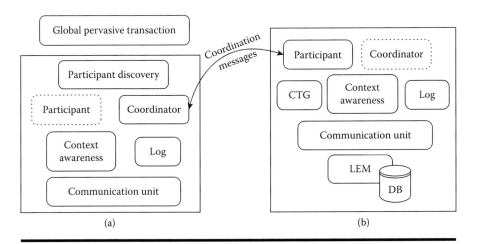

Figure 6.11 Architecture of the pervasive transaction system: (a) initiator and (b) executor.

service, and other 20 mobile devices (also executors) are able to handle public traffic queries. Fixed hosts support either sickbed reservations or public traffic queries. Any mobile device can issue global pervasive transactions, but only the executor is able to perform subtransactions initiated from other devices.

We simulated the mobility of nodes by changing link states among them, and furthermore, we let the wireless links disconnect in the probability *DisconnectProb*. In addition, we modeled system load in the number of concurrent pervasive transactions (simplified *NumMobiTran*). Pervasive transactions were randomly initiated and concurrently executed in the system. Each of them consisted of two subtransactions.

During the transaction processing, CATran dynamically discovers a qualified participant for a subtransaction T_i and can select the next subtransaction T_j substitutable for T_i if T_i was aborted. In the medical treatment reservation described in Section 6.1, for example, the coordinator will request the subtransaction T_5 after T_1 fails. By comparison, NonCATran dispatches subtransactions to well-known devices that provide the corresponding services so that a NonCATran transaction is aborted if the network link between an initiator and an executor is unqualified.

6.6.2 Results and Evaluation

High mobility and frequent network disconnection significantly decrease the probability of successful commits. As a result, the failed ratio (FR) is an important criterion for measuring the effectiveness and performance of a transaction management mechanism. FR refers to the fraction of failed transactions. It can be formulated as FR = (FN ÷ TN) × 100%, where FN is the number of failed transactions and TN is the total number of transactions in the system within a given period of time.

■ *System load.* In this experiment, the number of concurrent pervasive transactions is varied from 100 to 500, where link disconnection probability is fixed such that DisconnectProb = 0.1. The performance results obtained for the two kinds of transactions, CATran and NonCATran, are shown in Figure 6.12. From this figure, we can see that the FR of the system upgrades for both strategies as the transaction load increases and, for all ranges of the transaction load, CATran performs better than NonCATran. This is because CATran selects a qualified mobile device with qualified context and is also able to execute substitutable subtransaction(s) if the previous request failed. By comparison, NonCATran sends a transaction request directly, without context awareness. The reason for the increase in the FR with both execution strategies is that a greater load on the physical resources caused more heavy data conflicts.

■ *Link states.* When a link disconnects or does not have enough bandwidth, nodes connected with the link can no longer communicate with each other.

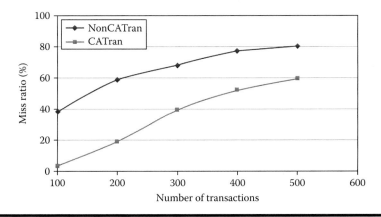

Figure 6.12 Failed ratio versus the number of concurrent pervasive transactions.

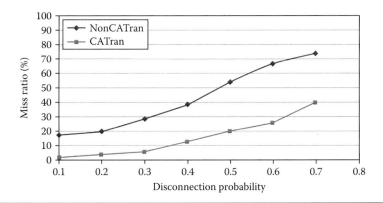

Figure 6.13 Failed ratio versus the disconnection probability of wireless links.

Therefore, for both CATran and NonCATran transactions, link states have significant impact on the FR. In this experiment, the *DisconnectProb is varied* from 0.1 to 0.7 in increments of 0.1. The performance results are shown in Figure 6.13, where the number of concurrent pervasive transactions in the system was fixed such that *NumMobiTran* = 50. As disconnection probability of wireless links increases, the performance of the system worsens with both execution strategies, CATran and NonCATran. The reason is that, with the increment of failure probability of links, more subtransactions cannot be sent to targeted nodes. However, the relative performance of CATran and NonCATran is not affected by the probability of wireless link failure because CATran transactions discover and select qualified devices before transaction processing.

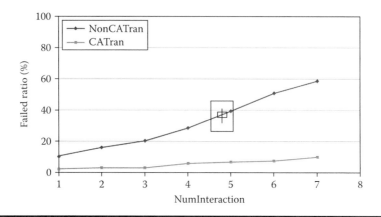

Figure 6.14 Failed ratio versus the number of interaction operations in a pervasive transaction.

■ *Interaction operations.* Interaction operations refer to the requests initiated by a mobile device during the execution of a transaction. In this experiment, the number of interaction operations in pervasive transactions is varied from 1 to 7 in increments of 1 to evaluate the impact of interaction operations. Performance results obtained are illustrated in Figure 6.14, where NumInteration means the number of interaction operations in a pervasive transaction. As the degree of interaction operations was increased, the performance of CATran gets only a little bit worse because the CATran transactions select qualified participants. By comparison, a steeper degradation in performance was observed in NonCATran strategy as the number of interaction operations increased. The reason is that NonCATran transactions are directly dispatched to a well-known device without selection so that failure probability of these transactions accumulated because of more interactions among an initiator and executors through unstable wireless connections.

Further Readings

Pervasive and Mobile Computing (Journal)
http://www.journals.elsevier.com/pervasive-and-mobile-computing/
Pervasive and Mobile Computing (PMC) is a professional, peer-reviewed journal that publishes high-quality scientific articles (both theory and practice) covering all aspects of pervasive computing and communications, such as wireless communications and networking, mobile computing and handheld devices, embedded systems, wearable computers, sensors, RFID tags, smart spaces, and middleware and software agents.

A Transaction Model for Mobile Computing
S. K. Madria and B. Bhargava, "A transaction model for mobile computing," *Database Engineering and Applications Symposium, 1998. Proceedings. IDEAS'98. International,* Cardiff, pp. 92–102, 1998.
This paper presents a transaction model for mobile computing. It introduces a prewrite operation before a write operation in a mobile transaction to improve data availability. The prewrite operation does not update the state of a data object but only makes visible the value that the data object will have after the commit of the transaction. This increases data availability during frequent disconnection common in mobile computing.

Neighborhood Consistent Transaction Management for Pervasive Computing Environments
F. Perich, A. Joshi, Y. Yesha, and T. Finin. In *Proceedings of the 14th International Conference on Database and Expert Systems Applications (DEXA 2003)*, 2003.
This paper examines the problem of transaction management in pervasive computing environments, presents a new approach for address them, and describes the implementation of the model and results from simulations.

A Survey of Academic and Commercial Approaches to Transaction Support in Mobile Computing Environments
C. Türker and G. Zini. *A Survey of Academic and Commercial Approaches to Transaction Support in Mobile Computing Environments.* Technical Report Number 429, ETH-Zentrum, CH-8092 Zürich, Switzerland, 2003.
Mobile transaction processing should be able to transparently deal with frequent disconnections and occasional movements of mobile devices. This paper surveys the mobile transaction models proposed in academia and outlines state-of-the-art transaction processing in commercial mobile databases.

References

1. T.G. Kanter. HotTown, enabling context-aware and extensible mobile interactive spaces. *IEEE Wireless Communications*, 9(5): 18–27, 2002.
2. A. Ranganathan, R.H. Campbell, A. Ravi, A. Mahajan. ConChat: A context-aware chat program. *IEEE Pervasive Computing*, 1(3): 51–57, 2002.
3. D. Bottazzi, R. Montanari, A. Toninelli. Context-aware middleware for anytime, anywhere social networks. *Intelligent Systems*, 22(5): 23–32, 2007.
4. K. Rehman, F. Stajano, G. Coulouris. An architecture for interactive context-aware applications. *IEEE Pervasive Computing*, 6(1): 73–80, 2007.
5. Z.W. Yu, X.S. Zhou, D.Q. Zhang, C.-Y. Chin, X. Wang, J. Men. Supporting context-aware media recommendations for smart phones. *IEEE Pervasive Computing*, 5(3): 68–75, 2006.
6. P. Heysters, G. Smit, E. Molenkamp. A flexible and energy-efficient coarse-grained reconfigurable architecture for mobile systems. *The Journal of Supercomputing*, 26(3): 283–308, 2003.
7. M. Liu, J.N. Cao, Y. Zheng. An energy-efficient protocol for data gathering and aggregation in wireless sensor networks. *The Journal of Supercomputing*, 43(2): 107–125, 2008.

8. M.H. Tu, P. Li, L.L. Xiao, I.-L. Yen, F.B. Bastani. Replica placement algorithms for mobile transaction systems. *IEEE Transactions on Knowledge and Data Engineering*, 18(7): 954–970, 2006.

9. E. Pitoura, P.K. Chrysanthis. Multiversion data broadcast. *IEEE Transactions on Computers*, 51(10): 1224–1230, 2002.

10. I. Chen, N. Phan, I. Yen. Algorithms for supporting disconnected write operations for wireless web access in mobile client-server environments. *IEEE Transactions on Mobile Computing*, 1(1): 46–58, 2002.

11. V. Lee, S. Son, E. Chan. On transaction processing with partial validation and time-stamp ordering in mobile broadcast environments. *IEEE Transactions on Computers*, 51(10): 1196–1211, 2002.

12. E. Pitoura, B.K. Bhargava. Data consistency in intermittently connected distributed systems. *IEEE Transactions on Knowledge and Data Engineering (TKDE)*, 11(6): 896–915, 1999.

13. J. Gray, P. Helland, P. O'Neil, D. Shasha. The dangers of replication and a solution. *ACM SIGMOD Record*, 25(2): 173–182, 1996.

14. M. Lee, S. Helal. HiCoMo: High commit mobile transactions. *Kluwer Academic Publishers Distributed and Parallel Databases (DAPD)*, 11(1): 73–92, 2002.

15. Q. Lu, M. Satynarayanan. Isolation-only transactions for mobile computing. *ACM Operating Systems Review*, 28(2): 81–87, 1994.

16. K.I. Ku, Y.S. Kim. Moflex transaction model for mobile heterogeneous multidatabase systems. In *Research Issues in Data Engineering (RIDE)*, pp. 39–46, 2000.

17. M.H. Dunham, A. Helal, S. Balakrishnan. A mobile transaction model that captures both the data and movement behavior. *Mobile Networks and Applications (MONET)*, 2(2): 149–162, 1997.

18. G.D. Walborn, P.K. Chrysanthis. Transaction processing in promotion. In *ACM Symposium on Applied Computing (SAC)*, pp. 389–398, 1999.

19. R.A. Dirckze, L. Gruenwald. A pre-serialization transaction management technique for mobile multidatabases. *Mobile Networks and Applications (MONET)*, 5(4): 311–321, 2000.

20. V. Kumar, N. Prabhu, M.H. Dunham, A.Y. Seydim. TCOT-a time-out-based mobile transaction commitment protocol. *IEEE Transactions on Computers*, 51(10): 1212–1218, 2002.

21. J.B. Lim, A.R. Hurson. Transaction processing in mobile, heterogeneous database systems. *IEEE Transactions on Knowledge and Data Engineering*, 14(6): 1330–1346, 2002.

22. D. Barbará. Mobile computing and databases—A survey. *IEEE Transactions on Knowledge and Data Engineering*, 11(1): 108–117, 1999.

23. J.L. Peterson. Petri nets. *Computing Surveys*, 9: 223–252, 1977.

24. T. Murata. Petri nets: Properties, analysis and applications. *Proceedings of the IEEE*, pp. 541–580, 1989.

Chapter 7

User Preferences and Recommendations

Pervasive applications assist human users to perform various tasks. In particular, these applications should become *smart* in the sense that even though the users may not explicitly specify their needs, the applications can learn from past interactions with users and adapt their behaviors in the future to provide customized services. In order to achieve this goal, pervasive applications actually learn user preferences from user interactions and make recommendations based on the learned preferences.

A recommendation is a way to help people find information to suit their needs and is widely used by many online services for suggesting products to customers that they might like to buy. For instance, Amazon uses an item-to-item collaborative filtering (CF) algorithm as their underlying recommender system [1]. Google uses CF to generate personalized recommendations for Google News users [2].

Previous recommender systems can be categorized into two types: *content-based approaches* [3] and *collaborative filtering* [4]. Content-based recommendations offer items to a user that are similar to past favorite. CF can be item-to-item (people who buy *X* also buy *Y*) or user-based (recommended by friends or users with similar interests) or hybrid approaches. Unlike content-based approaches, CF approaches make predictions from large-scale item–user matrices. Generally, CF can be classified into memory-based and model-based approaches. Memory-based schemes achieve high accuracy by exploiting similarity of functions between items and users [5–8]. Model-based CF approaches [1,9,10] first cluster all items or users into classes via machine learning algorithms and then use these classes for prediction. Recently, CF approaches based on matrix factorizations have been shown to be very effective for Netflix competitors [11,12].

This chapter is organized as follows. Section 7.1 discusses an RSS (*really simple syndication* or *rich site summary*) reader application that uses content-based recommendations. Then, we describe a CF algorithm that makes recommendations based on similar users and items in Section 7.2. Finally, Section 7.3 illustrates a top-K recommendation problem for social networks, where user preferences are computed as latent features, and a Monte Carlo simulation is utilized for making recommendations.

7.1 Content-Based Recommendation in an RSS Reader

This section discusses a content-based recommendation in an RSS reader, where recommendations are made according to user preferences captured during the use of the system. RSS is a technology that has been developed so that a site can create feeds for subscribed users to receive most recent updates. An RSS feed is an XML document containing information such as a title, full or summary text, publishing date, authorship, and URL, of new postings from a specific Web site. We call a posting in an RSS feed an item, which can be a piece of news, a blog, a tweet, or even a picture. To help users manage and read RSS feeds, many RSS readers [13] are developed that are capable of repeatedly fetching all RSS feeds and automatically checking for updates.

Currently, most RSS readers sort new items in the order of publishing date, which makes it hard for users to find information they like to read. The problem is that users may subscribe to a number of RSS feeds that may produce thousands of new items each day. As a result, it is difficult for users to find the most interesting items. The InterSynd system [14] recommends new RSS feeds (not new items) for users based on their neighbors' subscriptions. NectaRSS [15] is another RSS reader that builds a user's profile based on his browsing history and recommends items that are close to that profile—an approach based on text similarity. CRESDUP [3] also recommends advertisements for RSS feeds using text similarity. Our work differs from NectaRSS and CRESDUP in that we study a number of different features for RSS recommendation as well as the effectiveness of different features and feature combinations.

7.1.1 Data Collection

We have designed and implemented a Web-based RSS reader that uses YUI [16] to provide a more user-friendly Ajax-based user interface. Figure 7.1 shows the user interface of our RSS reader. In our RSS reader, users can add an RSS feed by providing the URL of the feed and then tagging on subscribe feeds. In the *all items* view, users can see a ranked list of items from all RSS feeds, with recommendations made by the system being displayed at the top of the page. After reading an item, users can click on a smile icon below the text of the item to mark the item as

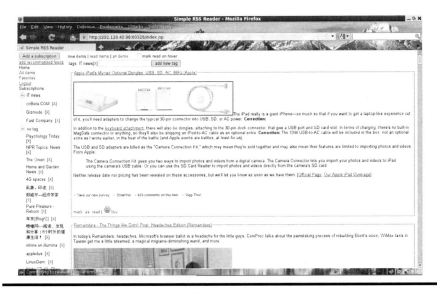

Figure 7.1   User interface of our RSS reader.

their favorite. In this way, we know the items that users like. To find new items, our system repeatedly fetches RSS feeds every 30 minutes so that it is unlikely to miss an item. Actually, 30 minutes is a very short time period; Google Reader checks for updates about once per hour [17].

When new items are downloaded, our RSS reader ranks them according to users' preferences and shows the ranked results for users. The idea is that users can see items they are interested in in top positions. Previous research [3,15] primarily has adopted text similarity for recommendations, but we also consider other features, such as users' past preferences for a feed (*favorite fraction*), update frequency of a feed, and PostRank [18] values. We study the effectiveness of recommendations for different features and feature combinations. The experimental results show that *favorite fraction* is the single most important feature, and the combination of text *similarity* and *favorite fraction* performs better than other combinations.

We invited 20 volunteers to use our Web-based RSS reader. Each time a user logs in, the reader creates a session and randomly chooses a method to rank all the new postings. The ranking methods include ordering by a single feature (publishing date, text similarity, and PostRank) and ordering by feature combinations. We only tell the volunteers that the system can recommend items for them, but they are unaware of any specific ranking method used in the session. In this way, they will not have bias toward any of the ranking methods. Each time a user marks an item as read or as his favorite, that item's information will be saved so that we can easily recalculate the ranking using other methods in the future.

In total, the system recorded 102 distinct sessions. Users read 6073 items using our RSS reader and marked 418 items as their favorites (about 6.9%). In sessions in which users read at least 100 items, the average number of favorite items was 7.8 per session.

7.1.2 Recommendation Features

We consider a number of features in our recommendation algorithms.

- *Similarity.* This feature represents the similarity between an item and the preference of a user. We follow the traditional information retrieval technique of using a vector space model to represent a text item or a user preference. Then, the similarity is calculated as the cosine value between two vectors.
- *Document vector.* The document vector of an item in an RSS feed is constructed with a TF-IDF weighting method [19]; that is,

$$w_d = \left(w_{t_1}, w_{t_2}, \ldots, w_{t_m} \right) \tag{7.1}$$

where w_d denotes the document vector for document d, and w_{t_i} denotes the TF-IDF value of term t_i in document d. Stop words are removed from the term vocabulary. Term frequency (TF) is calculated using maximum TF normalization (NTF):

$$ntf_{t,d} = \alpha + (1-\alpha) \frac{tf_{t,d}}{tf_{max}(d)} \tag{7.2}$$

where $tf_{t,d}$ is the number of occurrences when term t appears in document d, and $tf_{max}(d)$ is the maximum tf in document d. NTF is chosen over TF with the following considerations: The length of an RSS item has wide variations, resulting in large TF values for longer documents. This will give an unfair bias toward longer documents. On the other hand, NTF scales down TF values with maximum TF in the document and introduces parameter a for smoothing; thus, it is more effective for RSS feeds. The constant α in the equation is set to 0.5 as suggested in one study [19].

- *User preference.* A user's preference is constructed from items that have been marked as favorites by a certain user. We only consider the recent N favorite items of the user because a user's interests may vary over time. Specifically, a user's preference is defined as the centroid of his favorite items:

$$P_u = \frac{1}{N} \sum_{d \in F_u(N)} w_d \tag{7.3}$$

where P_u denotes the profile of user u, and $F_u(N)$ denotes the most recent N items that the user u marks as a favorite. We set N to 100 so that we have enough samples from the user and so that calculating the centroid vector is computationally efficient. The similarity between user profile and a new item (*Sim*) is calculated as their cosine distance:

$$Sim(P_u, w_d) = \frac{P_u \cdot w_d}{|P_u| \cdot |w_d|} \tag{7.4}$$

An item with high *Sim* is more relevant to a user profile, and the user may be interested in reading the item. Conversely, low *Sim* means irrelevant. *Sim* ranges from 0 to 1.

■ *Favorite fraction.* The favorite fraction (*FF*) feature measures the degree that a user may like an RSS feed. Specifically, the *FF* value for user u on feed i is defined as:

$$FF(i, u) = \frac{|S_{fav}(i, u)|}{|S_{read}(i, u)|} \tag{7.5}$$

where $|S_{fav}(i,u)|$ denotes the total number of items in feed i that user u has marked as his favorites, and $|S_{read}(i,u)|$ denotes the total number of items in feed i that user U has read. FF ranges from 0 to 1. *FF* shows the affection of a user to a specific feed. With a high *FF* value, the user will probably like new postings from the feed. For instance, a person may prefer *The New York Times* to the *Los Angeles Times* but subscribes to RSS feeds from both. With the FF feature, we can capture such a user preference on different RSS feeds.

■ *Inverse update frequency.* Update frequency measures the productivity of a feed. This is a feature solely related to RSS content providers. For many feeds, low productivity often leads to high quality. For example, the news RSS feeds from *Sina.com* publish hundreds of new items per day. It is unlikely that a user will be interested in all these new items. On the other hand, a blogger may only write a few blogs each day, and these blogs are of high quality to the user. Thus, we design an inverse update frequency (*IUF*) feature to capture the productivity of RSS feeds, which is defined as follows:

$$IUF(i) = 1 - \frac{n(i)}{n_{max}} \tag{7.6}$$

where n_{max} denotes the maximum number of updates per day of all the feeds, and $n(i)$ is the number of new items per day of feed i.

■ *PostRank.* The PostRank feature is obtained from *PostRank.com* [18], an online RSS reader. For each new RSS item, PostRank.com assigns a PostRank value (*PR*) according to the information about the item retrieved from other Web sites, such as the number of diggs and comments at *Digg.com*, how many people have bookmarked the item, whether or not the item is mentioned in Twitter or FriendFeed, and the number of clicks made by users using the PostRank Reader. A higher *PR* value means more people are talking about the item or more people like it. Thus, items with high *PR* are more likely to be of good quality or important. The original *PR* value ranges from 0 to 10, and we normalize it to be from 0 to 1.

7.1.3 Feature Combinations

The aforementioned features—text similarity, update frequency, favorite fraction, and PostRank—all have their merits as well as some shortcomings:

■ Text similarity may work well when a new item is close to the user preference—items that the user has marked as favorites before. However, similarity does not work well with new topics or new events that the user has never seen before or in cases where user interests have shifted.
■ *FF* always recommends items in feeds that users have read and liked in the past, but other feeds that user may not have read may be ignored.
■ High *IUF* or *PR* probably indicates items of high quality. But these two features have nothing to do with a user's preferences. In some cases, users may have little interest in some items with high *IUF* or high *PR* scores.

We consider ways of combining these features for better recommendations because these features present different aspects of a new item. Specifically, we adopt a linear combination of two features for ranking different items:

$$Rank = (1-k)f_1 + kf_2 \qquad (7.7)$$

where *Rank* is the final ranking score for an item; f_1 and f_2 can be any feature values of *Sim*, *IUF*, *FF*, and *PR*; and *k* is a constant value. For instance, we could set *k* to 0.3, f_1 to *Sim*, and f_2 to *FF*.

7.1.4 Performance

We assume that (1) users will like a new item no matter which position it is in and (2) users are focusing on reading while using the RSS reader. Thus, when a user marks certain items as favorites, we can expect that the user will like the same items using other ranking methods. In other words, we can re-rank items with different ranking algorithms and evaluate their performance with previously recorded

session data. To make sure that a user was concentrating on a session, we only perform re-ranking on sessions with more than 100 items. We use the following two
metrics in our evaluation:

- *Average favorite.* For sessions of more than 100 items, we re-rank all items in
  each session with different algorithms, and then we divide the first 100 items
  into five equal-sized bins—each bin consisting of 20 items. The average favorite (AF) of bin j for a specific ranking algorithm is defined as:

$$AF_j = \frac{\sum_u \sum_t fav(u,i,j)}{\sum_u \sum_t 1} \tag{7.8}$$

  where $fav(u,i,j)$ is the number of favorites of user u in session i and bin j.

  If, on average, most of the favorite items appear in bin 1, and few appear in
  other bins, then the ranking algorithm performs well. An algorithm performs
  poorly if favorite items in different bins are almost the same, or it performs
  even worse when favorites in the first bins are less than the latter bins.

- *Precision at N.* Precision at N measures the accuracy of the top N items for
  a specific ranking algorithm. We group all items of all sessions together and
  re-rank them with different algorithms. Given a number N, we calculate a
  specific algorithm's precision at N as:

$$Prec(N) = \frac{\text{number of favorites in top } N}{N} \tag{7.9}$$

  A ranking method works well if the precision at N is high.

- *Single feature.* This experiment studies the effectiveness of individual features
  for RSS recommendation. Figure 7.2 illustrates the average number of favorite items in different bins (i.e., the values of AF_j using different features).
  Among the four proposed features, FF and IUF perform better than Sim and
  PR. Specifically, FF has more than half (3.81) of the favorite items in the first
  bin, indicating that users do have a preference in RSS feeds. For Sim and IUF,
  both have decreasing AF values over bins. For ranking by age and PR, we can
  observe that the variation over five bins is not significant, indicating neither
  feature is effective in recommendations.

- *Feature combinations.* This experiment studies the effectiveness of different
  feature combinations. Table 7.1 illustrates the results of six-feature combination methods for the top 10 to top 50 items. We can observe that the combination of *similarity* and *favorite fraction* performs the best for all cases, with
  80% precision at the top 10. This is because both features are extracted from
  users' preferences. Combinations using *PostRank* generally perform poorly.

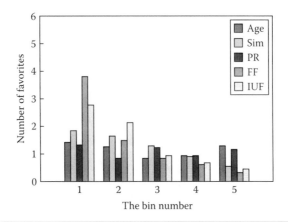

Figure 7.2   *AF* of ranking using different features.

Table 7.1   Precision at *N* for Different Feature Combination Methods of All Users (*k* = 0.7)

Top N	Sim+PR	Sim+IUF	Sim+FF	PR+IUF	PR+FF	IUF+FF
10	0.10000	0.30000	**0.80000**	0.00000	0.50000	0.60000
20	0.25000	0.25000	**0.75000**	0.00000	0.65000	0.70000
30	0.23333	0.46667	**0.53333**	0.00000	0.53333	0.53333
40	0.20000	0.50000	**0.55000**	0.00000	0.42500	0.42500
50	0.18000	0.50000	**0.50000**	0.00000	0.44000	0.44000

Note: Bold face values signify the best among all other features.

An extreme example is combining *PostRank* with *inverse update frequency*, which results in zero precision. This is because neither feature is related to user preference. Combinations using *favorite fraction* generally perform better than using other features, indicating it is an important feature for recommendations.

To summarize, we have proposed four different features for recommending new items to users in their subscribed RSS feeds. Our experimental results show that *favorite fraction* and *inverse update frequency* are effective and perform better than the simple text *similarity* between a new item and the preference of a user. On the other hand, *PostRank* and age-based ranking are ineffective for RSS recommendations. We also study the performance of combining two features together for recommendations. Our evaluation indicates that the combination of text *similarity* and *favorite fraction* performs the best.

7.2 A Collaborative Filtering-Based Recommendation

Unlike the previously mentioned content-based approach, CF makes predictions from large-scale item–user matrices and has achieved widespread success in recommender systems such as Amazon and Yahoo! music. However, CF usually suffers from two fundamental problems—data sparsity and limited scalability. Within the two broad classes of CF approaches—namely, memory-based and model-based—the former usually falls short of the system scalability demands because these approaches predict user preferences over the entire item-user matrix. The latter often achieves unsatisfactory accuracy because they cannot precisely capture the diversity in user rating styles.

We discuss an efficient CF approach using smoothing and fusing (CFSF). CFSF formulates a CF problem as a local prediction problem by mapping it from the entire large-scale item–user matrix to a locally reduced item–user matrix. Given an active item and a user, CFSF constructs a local item–user matrix as the basis of prediction. To alleviate the data sparsity, CFSF presents fusion strategy for the local item–user matrix, which fuses the ratings the same user gives to similar items and like–minded users give to the same as well as similar items. To eliminate the diversity in user rating styles, CFSF uses a smoothing strategy that clusters users over the entire item–user matrix and then smoothes ratings within each user cluster.

7.2.1 Background on Collaborative Filtering

Recommender systems aim at predicting the rating of active item i_a made by active user u_b from user profiles. These profiles are represented as a $Q \times P$ item–user matrix X, where Q and P are the sizes of X. The notations used are the following:

- $I = \{i_1, i_2,...,i_Q\}$ and $U = \{u_1, u_2,...,u_P\}$ represent the sets of items and users in X.
- $\{C_u^1, C_u^2,...,C_u^L\}$ are L user clusters, where users in each cluster share some similar tastes.
- $I\{u\}$, $I\{C_u\}$, and $U\{i\}$ denote the set of items rated by user u, the set of items rated by user cluster C_u, and the set of users who have rated item i.
- r_{u_b,i_a} denotes the score that user b rates item a, $\overline{r_{i_a}}$, and $\overline{r_{u_b}}$ represents the average ratings of item i_a and user u_b.
- Let SI, SU, and SUI be the sets of similar items, like-minded users, and similar items and like-minded users, respectively.
- SIR, SUR, and $SUIR$ denote predicting user preferences over the entire item–user matrix from the ratings the same user makes on similar items, the like-minded users make on the same item, and the like-minded users make on the similar items.
- SR represents predicting user preferences from all the ratings (i.e., SIR, SUR, and $SUIR$).
- Let SIR', SUR', $SUIR'$, and SR' be the counterparts of SIR, SUR, $SUIR$, and SR, but they are calculated over the local item–user matrix.

Then, the item vector of the matrix X is:

$$X_i = [i_1, i_2,..., i_Q],\ i_Q = [r_{1,q}, r_{2,q},..., r_{p,q}]^T$$

where $q \in [1,Q]$. Each column vector i_m corresponds to the ratings of a particular item m by P users. Matrix X can also be represented by user vectors illustrated as:

$$X_u = [u_1, u_2,..., u_p],\ u_p = [r_{p,1}, r_{p,2},..., r_{p,Q}]^T$$

where $p \in [1,P]$. Each row vector u_p^T indicates a user profile that represents a particular user's item ratings. Item-based CF approaches, as illustrated in Figure 7.3a, find the similar items among item vectors and then use the ratings made by the same user to predict his or her preferences. For example, given an active item i_a and a user u_b, Equation 7.10 denotes the mechanism of item-based CF approaches, where sim_{i_a,i_c} is the similarity of items i_a and i_c, and it is usually computed by Pearson correlation coefficient (PCC) or vector space similarity (VSS):

$$SIR:\ \widehat{r_{u_b,i_a}} \leftarrow \frac{\sum_{i_c \in SI} sim_{i_a,i_c} \cdot r_{u_b,i_c}}{\sum_{i_c \in SI} sim_{i_a,i_c}} \tag{7.10}$$

Alternatively, user-based CF approaches take advantage of similar motivation to predict user preferences, where the ratings of like-minded users for the active item are used. Equation 7.11 shows the mechanism of user-based CF approaches, where sim_{u_b,u_c} is the similarity of users u_b and u_c.

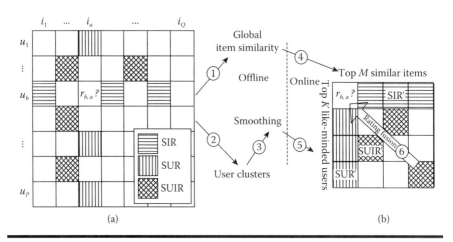

(a)   (b)

Figure 7.3 The CFSF approach. (a) the original item-user rating matrix; (b) the item-user rating matrix produced by CFSF for prediction.

$$SUR : \widehat{r_{u_b,i_a}} \leftarrow \frac{\sum_{u_c \in SU} sim_{u_b,u_c} \cdot r_{u_c,i_a}}{\sum_{u_c \in SU} sim_{u_b,u_c}} \tag{7.11}$$

Both item-based and user-based CF approaches do not consider *SUIR* for accuracy improvement. Let i be an item similar to i_a and u be a user that is like-minded to u_b; *SUIR* is calculated as:

$$SUIR : \widehat{r_{u_b,i_a}} \leftarrow \frac{\sum_{u,i \in SUI} sim_{(i,i_a),(u,u_b)} \cdot r_{u,i}}{\sum_{u,i \in SUI} sim_{(i,i_a),(u,u_b)}} \tag{7.12}$$

where $sim_{(i,i_a),(u,u_b)}$ is the weight for the rating user u makes on item i, denoting how much the rating $r_{u,i}$ is considered in prediction. The weight in CFSF is defined in Equation 7.21.

Finally, UI-based CF approaches [6,20] have proposed combining *SIR*, *SUR*, and *SUIR* with a fusion function. However, because searching for active items and users over the entire item–user matrix is time-consuming, all memory-based CF approaches achieve limited scalability.

7.2.2 CFSF Approach

7.2.2.1 Overview

Figure 7.3 explains how to derive CFSF from traditional CF approaches. First, CFSF creates a global item similarity (GIS) matrix performed in a memory-based manner (i.e., search over the entire item–user matrix). Then, CFSF classifies users into clusters, within each of which unrated ratings are smoothed. These two steps significantly reduce the influence of ratings diversity and accelerate the selection of like-minded users. Based on the request for an active user from the recommender system, CFSF picks up the top M similar items from GIS and the top K like-minded users from user clusters and extracts related ratings to create the local item–user matrix. In the end, it predicts user preferences by fusing the ratings *SIR′*, *SUR′*, and *SUIR′* over the local item–user matrix. Figure 7.3 also shows that CFSF significantly cuts down the number of users and items involved in prediction.

To cope with the increasing number of items and users in recommender systems, CFSF divides the CF process into offline and online phases. The offline phase involves three steps: creating the GIS, clustering users, and smoothing user ratings. This phase is a time-consuming process. The online phase includes the latter three steps: selecting items and users, constructing a local item–user matrix, and predicting preferences. Each user in CFSF has to rate a certain number of items so that recommendations for the user can be made.

7.2.2.2 Offline Phase

The three steps of the offline phase are:

1. *Creating GIS ← I.* In this step, CFSF creates *GIS* to save the item similarity matrix and to eliminate the diversity in item ratings. Items that are popular tend to get higher ratings than unpopular items. PCC, rather than pure cosine similarity (PCS), is selected as the item similarity function because PCS does not consider the diversity in item ratings. Given items i_a, i_b and $U = U\{i_a\} \cap U\{i_b\}$, the similarity between i_a and i_b is defined as:

$$sim_{i_a,i_b} = \frac{\sum_{u \in U}\left(r_{u,i_a} - \overline{r_{i_a}}\right)\cdot\left(r_{u,i_b} - \overline{r_{i_b}}\right)}{\sqrt{\sum_{u \in U}\left(r_{u,i_a} - \overline{r_{i_a}}\right)^2}\cdot\sqrt{\sum_{u \in U}\left(r_{u,i_b} - \overline{r_{i_b}}\right)^2}} \tag{7.13}$$

Given the large number of items, we set thresholds for Equation 7.13 to filter less important items. Then, the size of *GIS* will be greatly reduced.

2. *Clustering users C_i ← U.* In order to eliminate the diversity in user ratings, CFSF uses the *K*-means algorithm to cluster users and then smoothes ratings within each user cluster. The *K*-means method trains the data iteratively and assigns every user to a cluster whose centroid is closest to him or her. The time complexity of each iteration is linear in the size of the dataset. Compared with other clustering methods, *K*-means is simple, fast, and accurate. Its primary objective is minimizing $\sum_{i=1}^{k}\sum_{u_j \in c_j} sim|u_j - \overline{u}|$, where $\overline{u}$ is the centroid of all users that belong to C_i. Similarly, $sim|u_j - \overline{u}|$ is defined as Equation 7.14 based on a PCC similarity function, where u_a and u_b are users, and $I = I\{u_a\} \cap I\{u_b\}$, denoting an item set that both users u_a and u_b have rated:

$$sim_{u_a,u_b} = \frac{\sum_{i \in I}\left(r_{u_a,i} - \overline{r_{u_a}}\right)\cdot\left(r_{u_b,i} - \overline{r_{u_b}}\right)}{\sqrt{\sum_{u \in U}\left(r_{u_a,i} - \overline{r_{u_a}}\right)^2}\cdot\sqrt{\sum_{u \in U}\left(r_{u_b,i} - \overline{r_{u_b}}\right)^2}} \tag{7.14}$$

Thus, user clusters are generated for smoothing user ratings and for selecting the top *K* like-minded users.

3. *Smoothing user ratings within C_i.* There are two reasons for smoothing unrated data. First, users in the same cluster share similar tastes but have dissimilar rating styles; and so, the same item is rated differently by users belonging to the same cluster. This diversity in user ratings negatively affects the prediction accuracy. Second, rating data are quite sparse because users prefer not

to rate items and also cannot rate all items due to the overwhelming number of items in recommender systems. Consequently, the prediction accuracy is unsatisfactory without capturing such diversity. CFSF uses a smoothing strategy similar to SCBPCC [7] to fill unrated data. The smoothing function is defined as:

$$r_{u,i} = \begin{cases} r_{u,i}, & \text{if } u \text{ rates } i \\ \overline{r_u} + \Delta rC_{u',i}, & \text{otherwise} \end{cases} \tag{7.15}$$

where $\Delta rC_{u',i}$ is the deviation of the average rating of item i in $C_{u',i}$ that is a set of users who rate the item i in user cluster $C_{u'}$. $\Delta rC_{u',i}$ is given as Equation 7.16:

$$\Delta rC_{u',i} = \sum_{u \in C_{u',i}} \left(r_{u,i} - \overline{r_u}\right) / |C_{u',i}| \tag{7.16}$$

where $|C_{u',i}|$ is the size of $C_{u',i}$.

After smoothing, CFSF creates *iCluster* for each user to store its similarity to each user cluster. These *iClusters* are sorted in descending order and are used for selecting the top *K* like-minded users. As a result, they accelerate selection efficiency considerably. The feature of a user cluster is denoted as a centroid that represents an average rating over all users in the cluster. Given an item set $I = I\{u_a\} \cap I\{C_{u'}\}$, the similarity between user u_a and cluster $C_{u'}$ is defined as Equation 7.17. Thus, we get the *iCluster* for each user:

$$sim_{u_a, C_{u'}} = \frac{\sum_{i \in I} \left(r_{u_a, i} - \overline{r_{u_a}}\right) \cdot \Delta rC_{u',i}}{\sqrt{\sum_{i \in I} \Delta \left(rC_{u',i}\right)^2} \cdot \sqrt{\sum_{u \in U} \left(r_{u_a, i} - \overline{r_{u_a}}\right)^2}} \tag{7.17}$$

So far, we have described all the steps in the offline phase that are often computer-intensive, and hence, performed in the backend. The online phase of CFSF focuses on responding to requests, including constructing a local item–user matrix and fusing the ratings for prediction.

7.2.2.3 Online Phase

In general, user preferences are most likely derived from the most similar items and like-minded users. CFSF creates the local item–user matrix containing the most related users and items, and thus, yields significant savings in CPU time and memory cost. When a request comes, CFSF will pick up the top M similar items from *GIS*, the top K like-minded users from user clusters C, and will extract related ratings from the original item–user matrix.

- *Selecting top M similar items.* Recall that CFSF sorts the result as *GIS* in descending order when it creates a GIS matrix. Consequently, CFSF can directly pick up the top M similar items from *GIS*.
- *Selecting top K like-minded users.* User preferences are often scattered into several user clusters. Take movies, for example; user u may like action, fantasy, and crime films. To cover user preferences as much as possible, CFSF selects a user candidate set and then selects the top K like-minded users. To create a user candidate set, CFSF selects users from clusters in *iCluster* one by one. To select the top K like-minded users, CFSF varies the similarity function according to two types of ratings: original and smoothed ratings. It introduces a parameter w to differentiate these ratings. Given user u and active user u_a, the similarity function of selecting the top K like-minded users is defined as:

$$sim_{u_a,u} = \frac{\sum_f w_{u,i} \cdot \left(r_{u,i} - \bar{r}_u\right) \cdot \left(r_{u_a,i} - \bar{r}_{u_a}\right)}{\sqrt{\sum_f w_{u,i}^2 \left(r_{u,i} - \bar{r}_u\right)^2} \cdot \sqrt{\sum_f \left(r_{u_a,i} - \bar{r}_{u_a}\right)^2}} \quad (7.18)$$

where f denotes $i \in I\{u_a\}$ and w is the coefficient, defined as Equation 7.19. Depending on whether the rating is original or smoothed, the weighting coefficient w varies:

$$w : w_{u,i} = \begin{cases} \varepsilon, & \text{if } u \text{ rates } i \\ 1-\varepsilon, & \text{otherwise} \end{cases} \quad (7.19)$$

Compared with previous methods, CFSF reduces the computation overhead by selecting the like-minded users from iCluster rather than from the entire item–user matrix. Moreover, CFSF is capable of setting thresholds for Equation 7.18, which will further reduce the computation overhead. After the top M similar items and the top K like-minded users are selected, CFSF will extract related ratings from the original item–user matrix to fill the local item–user matrix.

- *Fusing SIR′, SUR′, and SUIR′.* CFSF defines *SIR′*, *SUR′*, and *SUIR′* as predictions of user preferences over the local item–user matrix from the ratings the same user makes on similar items and the like-minded users make on the same and similar items. M and K are much less than Q and P. As a result, *SIR′*,

SUR', and *SUIR'* are computed much faster than *SIR*, *SUR*, and *SUIR*, which are calculated over the entire large-scale item–user matrix. Given active item i_a and user u_b, Equation 7.20 illustrates these definitions:

$$SIR' = \frac{\sum_{s=1}^{M} w \cdot sim_{i_s,i_a} \cdot r_{u_b,i_s}}{\sum_{s=1}^{M} w \cdot sim_{i_s,i_a}}$$

$$SUR' = \frac{\sum_{t=1}^{K} w \cdot sim_{u_t,u_b} \cdot \left(r_{u_t,i_a} - \overline{r_{u_t}}\right)}{\sum_{t=1}^{K} w \cdot sim_{u_t,u_b}} + \overline{r_{u_b}} \qquad (7.20)$$

$$SUIR' = \frac{\sum_{K}^{K} \sum_{s=1}^{M} w \cdot sim_{(i_s,i_a),(u_t,u_b)} \cdot r_{u,i}}{\sum_{t=1}^{K} \sum_{s=1}^{M} w \cdot sim_{(i_s,i_a),(u_t,u_b)}}$$

where w is defined as Equation 7.19 and $sim_{(i,i_a)(u,u_b)}$ is defined by Euclidean distance as Equation 7.21, denoting the weight for the rating of the similar item i by the like-minded user u.

$$sim_{(i_s,i_a)(u_t,u_b)} = \frac{sim_{i_s,i_a} \cdot sim_{u_t,u_b}}{\sqrt{sim_{i_s,i_a}^2 + sim_{u_t,u_b}^2}} \qquad (7.21)$$

CFSF selects *SUR'* as the major prediction tool and *SIR'* and *SUIR'* as supplementary when it predicts user preferences. CFSF introduces two parameters, λ and δ, to balance the impact of *SIR'*, *SUR'*, and *SUIR'*. The fusion function of CFSF is defined as:

$$SR' : \widehat{r_{u_b,i_a}} = (1-\delta) \cdot (1-\lambda) \cdot SIR' + (1-\delta) \cdot \lambda \cdot SUR' + \delta \cdot SUIR' \qquad (7.22)$$

where λ and δ are between 0 and 1.

To summarize, the offline phase of CFSF has a high cost due to the creation of the GIS matrix and clustering. To reduce computation overhead in the offline phase, CFSF sets thresholds to filter items. In the online phase, the time complexity of CFSF is $O(MK)$, where M and K are the number of similar items and like-minded users. Considering that M and K are much less than the original sizes of an item–user matrix, the online phase of CFSF is scalable.

7.2.3 Performance of CFSF

We evaluated the performance of CFSF with the MovieLens dataset [21] from the University of Minnesota. We randomly extracted 500 users from MovieLens, where each user rated at least 40 movies. We changed the size of the training set by selecting the first 100, 200, and 300 users, denoted as ML_100, ML_200, and ML_300, respectively. We selected the last 200 users as the test set. We varied the number of items rated by active users from 5 to 10 to 20, denoted as Given5, Given10, and Given20, respectively. Table 7.2 summarizes the statistical features of the datasets used in our experiments.

7.2.3.1 Metrics

To be consistent with experiments reported in the literature [5–7,22,23], we use the same MAE as the evaluation metric, defined as:

$$MAE = \frac{\sum_{u \in T} \left| r_{u,i} - \overline{r_{u,i}} \right|}{|T|} \tag{7.23}$$

where $r_{u,i}$ denotes the rating that user u gives item i, T represents the test set, and $|T|$ is the test size. The smaller the value of MAE, the better the performance.

7.2.3.2 Overall Performance

We carried out experiments from two aspects to evaluate the performance of CFSF. One aspect is to compare CFSF with traditional memory-based CF approaches: an item-based approach using PCC (SIR) and a user-based approach using PCC (SUR). For MovieLens, the parameters of CFSF are set as follows: $C = 30$, $\lambda = 0.8$, $\delta = 0.1$, $K = 25$, $M = 95$, and $w = 0.35$. Table 7.3 illustrates the results, showing that CFSF considerably outperforms the SUR and SIR with respect to prediction accuracy.

We compare CFSF with the other state-of-the-art CF approaches; that is, AM [24], EMDP [20], PD [25], SCBPCC [7] and SF [6]. We varied the item number that each user was required to rate on all the test sets for the MovieLens dataset. The results are

Table 7.2 Statistics of the Datasets

	MovieLens
No. of users	500
No. of items	1000
Average no. of rated items per user	94.4
Density of data	9.44%
No. of ratings	5

Table 7.3 MAE on MovieLens for the SIR, SUR, and CFSF

Training Set	Methods	Given5	Given10	Given20
ML_300	CFSF	**0.743**	**0.721**	**0.705**
	SUR	0.838	0.814	0.802
	SIR	0.870	0.838	0.813
ML_200	CFSF	**0.769**	**0.734**	**0.713**
	SUR	0.843	0.822	0.807
	SIR	0.855	0.834	0.812
ML_100	CFSF	**0.781**	**0.758**	**0.746**
	SUR	0.876	0.847	0.811
	SIR	0.890	0.801	0.824

Note: Bold face values signify the best MAE among all methods.

Table 7.4 MAE on MovieLens for the State-of-the-Art CF Approaches

Training Set	Methods	Given5	Given10	Given20
ML_300	CFSF	**0.743**	**0.721**	**0.705**
	AM	0.820	0.822	0.796
	EMDP	0.788	0.754	0.746
	SCBPCC	0.822	0.810	0.778
	SF	0.804	0.761	0.769
	PD	0.827	0.815	0.789
ML_200	CFSF	**0.769**	**0.734**	**0.713**
	AM	0.849	0.837	0.815
	EMDP	0.793	0.760	0.751
	SCBPCC	0.831	0.813	0.784
	SF	0.827	0.773	0.783
	PD	0.836	0.815	0.792
ML_100	CFSF	**0.781**	**0.758**	**0.746**
	AM	0.963	0.922	0.887
	EMDP	0.807	0.769	0.765
	SCBPCC	0.848	0.819	0.789
	SF	0.847	0.774	0.792
	PD	0.849	0.817	0.808

Note: Bold face values signify the best MAE among all methods.

shown in Table 7.4. As the test set increases, the MAEs of all approaches show a downward trend. As the number of rated items for each user increases from 5 to 20, a similar trend is observed. Among them, CFSF achieves the best accuracy. This is because CFSF can select the most similar items and like-minded users.

7.3 Preference-Based Top-*K* Recommendation in Social Networks

Social networks are social structures representing the relationships among people. Examples of social networks include Facebook online, mobile phone networks [26], and scientific collaboration networks [27]. A social network is an important way of spreading information and works via the *word-of-mouth* or *viral marketing* approach [28,29]. The selected users influence their friends on the social network, and friends influence their friends. Finally, a large number of users would choose the product. Besides marketing, other important applications on the social network include finding the most important blogs [30] and searching for domain experts [31,32].

All of the aforementioned applications can be generalized as the problem of finding top-*K* influential nodes in networks (i.e., which *K* users should be selected so that they eventually influence the most people in the network). More formally, this can be formulated as a discrete optimization problem called an *influence maximization problem* [33]; for a parameter *K* and a social network, where influence is propagated in it according to a stochastic cascade model, find a *K*-node set that eventually has the maximum influence.

The influence maximization problem was first studied as viral marketing strategies [34,35] and revenue maximization [28]. Kempe et al. [33] formulate influence maximization as a discrete optimization problem and prove it to be NP-hard. They present a greedy approximation algorithm and prove that the optimal solution for an influence maximization problem can be approximated to within a factor of $(1-1/e)$.

The algorithm of Kempe et al. is inefficient and more efficient algorithms [26,27,30,36] have been proposed, which have been extended to social networks at a greater scale. CELF [30] optimizes the greedy algorithm by exploiting the submodularity property to reduce the number of evaluations on the influence spread of users, which we borrowed to use in this work. NewGreedy [36] removes edges that will not contribute to propagation to get a smaller graph. MixedGreedy [36] combines the CELF and the NewGreedy, where the first round uses NewGreedy and the rest of the rounds use CELF. CGA [26] first detects communities in a social network and then employs a dynamic programming algorithm for selecting communities to find influential users. Kimura and Saito [27] propose shortest-path-based influence cascade models and provide efficient algorithms for computing influence spread under these models. However, they do not directly address the efficiency issue of the greedy algorithms studied by Kempe et al. because the influence of cascade models is different.

All of these approaches do not consider user preferences during influence diffusion and use a uniform probability. However, some studies [37,38] have found that the diffusion of different topics in the network is not the same because users have different preferences that will affect their roles in the spread of a specific topic [39]. Unfortunately, although Tang et al. [39] combine the specific topic and social influence analysis together, demonstrating that different topics lead to different influence results, there is no research about an influence maximization problem taking the user preferences into account. This drawback greatly affects the accuracy of greedy approximation algorithms proposed by Kempe et al. and others.

We study the influence maximization problem considering user preferences. We propose a two-stage algorithm, called the greedy algorithm based on users' references (GAUP), for mining top-K influential users for a given topic. The first stage calculates each user's preference for a specific topic by singular value decomposition (SVD)–based latent semantic indexing (LSI) [40]. The second stage combines the traditional greedy algorithms and user preferences calculated from the first stage, and then it computes an approximate solution for the influence maximization problem for a specific topic. Our evaluation results with an academic social network demonstrate that our GAUP algorithm can maximize the influence spread on a topic. We have compared GAUP with an SVD-based CF algorithm and HITS for expert search, and we have found that GAUP is more likely to find the most influential domain experts than CF. In addition, GAUP is more reliable than HITS because HITS suffers from the problem of topic drift.

7.3.1 Problem Formulation

We formulate the problem as follows. A social network is an undirected graph $G = (V,E)$. Vertices in V are the nodes in the network, and edges in E model the relationship between nodes. There is a set of users' documents with topic labels. Given a number K and a topic label T, the task is to generate a seed set S of K vertices, with the objective that influence spread is as large as possible for topic T from the seed set.

We extract an academic coauthor network from DBLP, where G is a coauthorship graph, vertices are authors of academic papers, and an edge indicates that the corresponding two authors have coauthored a paper. Parallel edges between two vertices denote the number of papers coauthored by the two authors. In DBLP, we can get information on each paper, including authors and conferences. We regard each conference as a distinct topic and each paper as a labeled document. Table 7.5 lists the notations used in this section.

When considering models for influence spread through a social network, we often say a node is either *active* or *inactive*. Active nodes are those that are influenced by other active nodes and that are able to influence their inactive neighbors when they become active; inactive nodes are those that have not been influenced by their active neighbors. The influence diffusion from the perspective of an initially inactive node v unfolds as follows: As time progresses, more and more of v's

Table 7.5   Notations

Terms	Descriptions
$G = (V,E)$	A graph G with vertex set V and edge set E
K	Number of seeds to be selected
R	Number of rounds of simulations
T	A selected topic (conference)
p	The uniform propagation probability
$p_{v,w}$	The propagation probability that node v activates node w
$C_{a,T}$	Node a's preference on topic T
S	The seed set of K influential nodes
$IS(S)$	Influence spread of S
$ISST(S,T)$	Influence spread of S on a specific topic T

neighboring nodes become active; v may become active at some point because of this; and v's activation may also affect its neighboring nodes. Each node can switch from an inactive to active state but not the opposite.

Popular influence diffusion models include the *linear threshold model* and the *independent cascade model* (IC). In the linear threshold model, a node v becomes active if all its neighboring nodes' influence added together exceeds a threshold. All nodes that were active in step $t-1$ remain active in step t. The threshold for each node is uniformly distributed within interval [0,1], representing the fraction of v's neighbors that must be active in order to activate v.

In the IC model, the diffusion process also unfolds in discrete steps; when an active node v becomes active in step $t-1$, it is given a single chance to activate each of its inactive neighbors with a probability p, which is independent of the history thus far. If multiple nodes try to activate the same node, their attempts are sequenced in an arbitrary order. If successful, the activated node becomes active in step t. Whether or not v succeeds, it cannot make any further attempts to activate its neighboring nodes in subsequent rounds. The process ends when no activations are possible.

The activation probability p in the work of Kempe et al. is uniform to each edge of the graph. If node v and w have $c_{v,w}$ parallel edges, v has a total probability of $1-\left(1-p\right)^{c_{v,w}}$ to activate w once it becomes active.

These models do not consider topics and users' preferences on different topics. Intuitively, the more preferences on topic T by v and its neighboring nodes, the more likely the influence will happen. Thus, we consider the activation probability p is not uniform but is related to users' preferences on a topic T.

7.3.2 Computing User Preferences

To compute user preferences for a specific topic, we adopt the SVD-based LSI model that can project user preferences into a reduced latent space. By assigning a weight for each latent item, our algorithm can compute the preference value of a user for a given topic.

SVD [41,42] is a well-known matrix factorization method that factors an $n \times c$ matrix R into three matrices:

$$R = U \cdot \Sigma \cdot V^T \tag{7.24}$$

where U and V are two orthogonal matrices of size $n \times n$ and $c \times c$, respectively. Σ is a diagonal matrix of size $n \times c$ that has all singular values of matrix R as its diagonal entries. The rank of R and Σ is r ($r \leq c \leq n$). All the entries of matrix Σ are positive and are stored in decreasing order of their magnitude.

Typical usage of SVD keeps k largest diagonal values of Σ, thus producing a rank-k approximation matrix R_k. Over all rank-k matrices, R_k minimizes the Frobenius norm $\|R - R_k\|$. In other words, SVD provides the best lower rank approximations of the original matrix R. The selection of k should be large enough to capture all the important structures in the original matrix and small enough to avoid overfitting errors.

We use SVD to compute user preferences, which are represented as low-dimensional latent features learned from a user–topic matrix. Specifically for a given user–topic matrix M, our algorithm computes preference prediction using the following steps:

1. Factor M using SVD to obtain U, Σ, and V.
2. Reduce the matrix Σ to dimension k.
3. Compute the square root of the reduced matrix Σ_k to obtain $\Sigma_k^{1/2}$.
4. Compute two matrices X and Y, where $X = U_k \Sigma_k^{1/2}$ and $Y = \Sigma_k^{1/2} V_k^T$.
5. Predict user a's preference on topic T as follows:

$$C_{a,T} = C_0 + X(a) \cdot Y(T) \tag{7.25}$$

X is a matrix with dimension of $n \times k$, describing the authors' preferences in the k-dimensional latent space (i.e., the weights of k latent features by the authors). $X(a)$ denotes ath row of X, representing user a's weights for k latent features. Y is a matrix with dimension of $k \times c$, representing the relationship between k-dimensional latent space and c topics. $Y(T)$ denotes Tth column of Y, representing k latent features' weights to Tth topic. C_0 is a constant.

To compute the predictions, we simply calculate the dot product of the ath row of X and the Tth column of Y and add the C_0 back. $C_{a,T}$ is the prediction of ath author's preference for topic T.

Table 7.6 Translation Rules for Converting Publication Counts to 1–5 Ratings

# in this conference/personal total #	Ratings
0	0
(0, 0.02]	1
(0.02, 0.05]	2
(0.05, 0.1]	3
(0.1, 0.15]	4
(0.15, 1)	5

From the DBLP dataset, we can get a user–topic matrix M, where rows are authors and columns are conferences. The question is how to define values in matrix M. Inspired by the rating matrix used in CF, entries in matrix M are rated with a five-star scale, where high scores mean that users prefer the conference more favorably. Table 7.6 lists rules for such a translation.

Following these rules, we obtain the user–topic matrix M, where M_{ij} denotes the preference of author i for conference j. A zero value means author i is not interested in conference j or has not published in the conference.

The resulting matrix is large and extremely sparse. Performing SVD on such a matrix is computationally intensive. To solve this problem, we remove authors who have published in less than 10 conferences from the matrix, thus reducing the size of the matrix significantly. The intuition is that we are interested in the most influential people among all authors, and these important ones tend to publish heavily at many different conferences.

7.3.3 Greedy Algorithm for Mining Top-K Nodes

In the traditional IC model, the probability of an activation is uniform to all edges. However, such a uniform probability cannot capture the preferences of users on different topics. For mobile social networks, Wang et al. [26] extend the IC model to accommodate contact frequency. Similarly, GAUP considers that the probability of influence is not only associated with influence frequency but also correlated with user preferences: the more preferences on topic T by v and w, the more likely the influence will happen. In order to describe this property, we define $p_{v,w}$ as follows:

$$p_{v,w} = p \cdot \mathcal{F}\left(C_{v,T}, C_{w,T}\right) \tag{7.26}$$

This new influence model is called the extended independent cascade (EIC) model. In the aforementioned formula, the original uniform probability p is weighted with a user preference function $\mathcal{F}(\cdot)$. The $C_{v,T}$ and $C_{w,T}$ are calculated by Equation 7.25. In experiments, we choose the function $\mathcal{F}(C_{v,T},C_{w,T})$ to be the square of the product of $C_{v,T}$ and $C_{w,T}$. When node v and w have $C_{v,w}$ parallel edges, v has a total probability of $1-(1-p_{v,w})^{c_{v,w}}$ to activate w once it becomes active.

Mining top-K influential nodes in the new EIC model is NP-hard but can be approximated with a greedy hill-climbing algorithm. Nemhauser et al. [43,44] have shown that if the influence function $f(\cdot)$ is submodular, nonnegative, and monotone, the NP-hard *influence maximization problem* can be approximated by a greedy hill-climbing algorithm within a factor of $(1-1/e)$. The EIC model is an edge-weighted version of an IC model, for which Kempe et al. [33] prove that the influence function is submodular. The objective function is obviously non-negative and monotone. Thus, we can use the greedy strategy to obtain a $(1-1/e-\varepsilon)$ approximation [33].

To compute the influence spread in mining the seed set and in evaluating the performance of algorithms, we often use Monte Carlo simulations. Let S be the seed set computed from the greedy algorithm and $RS(S)$ denote the resulting set of influence cascade. The influence spread after the random process is $|RS(S)|$. For any topic T, we obtain a user-preference vector $UP(T)$ from the previous step. GAUP takes the graph G, the number K, and the vector $UP(T)$ as input and generates a seed set S of cardinality K, with the purpose that the expected influence spread by the seed set S is as large as possible for topic T.

The idea of the greedy algorithm is to run for K rounds, where each round finds the vertex that will maximize the incremental influence spread in the round. Algorithm 7.1 describes the details of GAUP. First, we calculate the propagation probability of each edge in the social network using the vector $UP(T)$, based on the EIC model (lines 5–8). Then, for each vertex, we compute its influence spread and insert the vertex into a priority queue, sorted by influence spread values (lines 10–17). The first vertex chosen will be the head element in the queue (lines 19–21). For the rest $K-1$ vertices, the algorithm first pops a vertex from the priority queue. If the vertex has been visited in this round, then we found the vertex for this round (lines 25–30). Otherwise, the increase of influence spread by adding this node to current set S is computed, and the vertex is inserted back to the priority queue (lines 32–37).

Note that lines 23–39 implement the CELF optimization [30], which takes advantage of the submodularity property of the influence maximization objective. By exploiting submodularity during each round, the incremental influence spread of a large number of nodes does not need to be re-evaluated when their values in the previous round are less than the values of some other nodes in the current round. Previous work [30] has shown that this optimization can improve the performance by a factor of up to 700 times.

Algorithm 7.1   The Greedy Algorithm for Mining Top-*K* Influential Nodes

1: **Input:** graph $G = (V, E)$, number K, topic T, user preference vector $U P (T)$
2: **Output:** a seed set S
3:
4: Initialize $S = \varnothing$, $R = 10000$, $Q = \varnothing$, $current\_is = 0$
5: // Compute activation probability of each edge
6: **for** each edge $(v, w) \in E$ **do**
7: $p_{v,w} = p \cdot F (C_v, T, C_w, T)$
8: **end for**
9:
10: // Compute influence spread of each vertex
11: **for** each vertex $v \in V$ **do**
12: $is_v = 0$
13: **for** $j = 1$ to R **do**
14: $is_v += |RS(v)|/R$
15: **end for**
16: insert $node(v, is_v)$ to Q
17: **end for**
18:
19: // 1st vertex is the one that has the largest influence spread
20: $node = pop(Q)$
21: $S = \{node.v\}$, $current\_is = node.is$
22:
23: **for** $i = 2$ to K **do**
24: **while** true **do**
25: $node = pop(Q)$
26: **if** $node.v$ has been visited in this round i **then**
27: $S = S \cup \{node.v\}$

(Continued)

Algorithm 7.1 *(Continued)* **The Greedy Algorithm for Mining Top-*K* Influential Nodes**

28:	*current is+ = node.is*
29:	**break**
30:	**end if**
31:	
32:	*v = node.v, is = 0*
33:	**for** *j* = 1 to *R* **do**
34:	*is+ = |RS(S ∪ {v}|)|*
35:	**end for**
36:	*is = is/R − current is*
37:	insert *node(v, is)* to *Q*
38: **end while**	
39: **end for**	

7.3.4 Performance

We conduct experiments on an academic coauthorship dataset from DBLP, which provides bibliographic information on major computer science journals and proceedings. DBLP indexes more than one million articles. In the extracted coauthor network, each node represents an author, and an edge represents that the corresponding two authors have collaborated once. Parallel edges mean that the authors have collaborated more than once. We only select the papers published in conferences and extract the authors who have published in at least 10 conferences. The generated network with papers from 411 conferences contains 8627 nodes and 91,574 edges.

7.3.4.1 Influence Model

We conduct experiments to compare the state-of-the-art greedy algorithm using an IC model and the GAUP using an EIC model. In the IC model, the probability p is set to 0.01 [33]. In the EIC model, the probability $p_{v,w}$ is computed by Equation 7.2.

7.3.4.2 Algorithms and Metrics

Table 7.7 lists the algorithms that are compared within the experiments. GA and GAUP are greedy algorithms mentioned in Section 7.3.3. CF is the collaborative

Table 7.7 The Algorithms

Algorithms	Descriptions
GAUP	Greedy algorithm using LSI-based user preferences
GA	The state-of-the-art greedy algorithm
CF	Collaborative filtering algorithm based on SVD
Random	Baseline of choosing random nodes in the graph
HITS	Hyperlink-induced topic search algorithm [44]

filtering algorithm based on SVD that selects nodes with biggest $C_{v,T}$ values as the seed set. For an expert search, we also consider the HITS algorithm [45]. For coauthors of a paper, we add links pointing to the first author from all other coauthors.

For evaluation, the following metrics are used:

■ ISST(S,T): Influence spread of seed set S on a specific topic T. This metric evaluates the influence on a specific topic that takes into account user preferences. ISST is defined as Equation 7.2, where node v belongs to all nodes being activated by seed set S.

$$ISST(S,T) = \sum_{v \in RS(S)} C_{v,T}^2 \qquad (7.27)$$

■ IS(S): The traditional influence spread of seed set S that does not take user preferences into account.

To obtain the IS(S) and ISST(S,T) for each seed set S, we run *Monte Carlo* simulations (using the corresponding IC and EIC model) 10,000 times and take the average. For all these algorithms, we compare their IS(S) and ISST(S,T) with different S sizes of 5, 10, 15, and 20.

7.3.4.3 Study on Influence Spread

We evaluate the performance of the proposed GAUP algorithm in a real coauthor network and compare GAUP with other algorithms on the specific topic of *SIGCOMM*. Figure 7.4 shows the ISST of various algorithms using the EIC model. From the figure, we can observe that *random* as the baseline performs very badly. GAUP performs significantly better than other algorithms when $K \geq 15$. GAUP outperforms GA by about 10% when K is 15 and more than 30% when K is 20. This is because GAUP can find more important nodes with respect to the specified topic. When K is 5 or 10, GA finds nodes that can activate more nodes

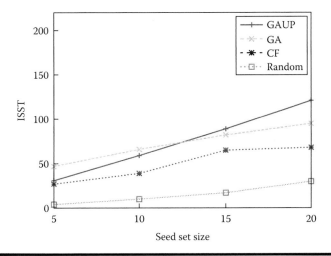

Figure 7.4   A comparison of ISST on a specific topic.

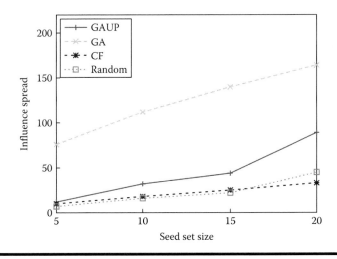

Figure 7.5   A comparison of influence spread of different algorithms.

than GAUP (shown in Figure 7.5). As a result, the cumulative influence of GA on the topic exceeds that of GAUP. CF performs worse than GA by about 30% because CF does not consider the influence diffusion process.

Figure 7.5 compares the IS(S) of various algorithms in the IC model. IS(S) does not take user preferences into account. As a result, the performance of GAUP is not as good as GA because GA can find most influential nodes across all topics. GAUP performs significantly better than *random* and CF because GAUP can find most influential nodes in a specific topic. CF does not perform well because the influence diffusion process is not used.

7.3.4.4 ISST versus IS

We study the effectiveness of ISST and IS when considering users' topic preferences. In the experiment, we first use the GAUP algorithm to generate two seed sets, *comm* and *kdd*, for topic *SIGCOMM* and *SIGKDD*, respectively. With these two seed sets, we compute their IS and ISST values. Table 7.8 shows that when K increases, IS(*comm*) and IS(*kdd*) are becoming very close. In contrast, Table 7.9 illustrates the ratios of ISST of *comm* over ISST of *kdd* for different topics. For topic *SIGCOMM*, we can observe that ISST(*comm,SIGCOMM*) is significantly larger than ISST(*kdd,SIGCOMM*), more than two times when $K\geq15$. On the other hand, ISST(*comm,SIGKDD*) is only about one-third of ISST(*kdd,SIGKDD*) for topic *SIGKDD* when $K\geq15$. These results demonstrate that ISST effectively measures the influence spread on specific topics and should be used as the metric other than the traditional IS when considering users' preferences.

7.3.4.5 Expert Search

This experiment studies whether the GAUP algorithm can find experts in a specific domain. Again, we choose *SIGCOMM* as the topic. Table 7.10 shows the seed sets obtained by GA, CF, GAUP, and HITS when $K = 5$. None of the authors selected by GA is in the networking field because GA does not take domain preference into account and tends to find experts in other domains. In contrast, GAUP, CF, and HITS can successfully find networking experts. We verified that all these people have published in well-known networking conferences. Except for Jennifer Rexford, all the other four experts selected by GAUP and CF are different. James F. Kurose appears in both GAUP and HITS. Joseph Naor publishes lots of papers at networking conferences but none in *SIGCOMM*. This indicates that our algorithm can effectively find domain experts.

Table 7.8   IS(comm) and IS(kdd)

Seed Set	5	10	15	20
comm	11	30	47	86
kdd	30	41	56	89

Table 7.9   Ratios of ISST of comm over ISST of kdd for Different Topics

Topic	5	10	15	20
SIGCOMM	1.67	2.47	2.91	2.81
SIGKDD	0.13	0.16	0.31	0.27

Table 7.10   The Seed Set of GA, CF, and GAUP When *k* = 5

GAUP	GA	CF	HITS
James F. Kurose	Wei-Ying Ma	Simon S. Lam	Donald F. Towsley
Jennifer Rexford	Philip S. Yu	Jennifer Rexford	James F. Kurose
Joseph Naor	Wen Gao	Vishal Misra	Zhen Liu
Deborah Estrin	Mahmut T. Kandemir	Donald F. Towsley	Weibo Gong
Thomas E. Anderson	Thomas S. Huang	Lixia Zhang	Vishal Misra

To further study the effectiveness of GAUP, CF, and HITS, we designed the following experiments to compare GAUP and CF and to demonstrate the topic drift problem of HITS.

7.3.4.6 GAUP versus CF

The GAUP and CF results given in Table 7.10 generally make it hard to measure which set of users is better. To address this problem, we design the following experiment. For the field of networks, we choose two topics, *SIGCOMM* and *ICCCN*. Then GAUP and CF algorithms are run to find the top 5, 10, 15, and 20 experts. Finally, we consider experts that are found by both topics.

Figure 7.6 illustrates the results of these two algorithms. We can observe that for the top 5, 10, and 15 experts, CF finds none that appear in both topics. In contrast, GAUP discovers 2, 4, and 5 overlapping ones for the top 5, 10, and 15, respectively. CF only finds one overlapping for the top 20 users, in which case GAUP finds 7. This experiment shows that GAUP can find common experts for different subtopics in a domain. Thus, these experts are more likely to be the most influential ones in the domain. The CF algorithm does not consider the influence spread process and only finds experts for specific topics. As a result, CF does not produce results as good as those of GAUP.

7.3.4.7 Topic Drift of HITS

The effectiveness of HITS depends on the quality of the initial set of nodes. Previous work [46,47] has shown that HITS may suffer from topic drift. To illustrate this problem, we run the HITS algorithm with three different seed sets for topic *SIGCOMM*:

■ *20-set.* We select 20 users from the experts returned by GAUP and CF for topic *SIGCOMM*.

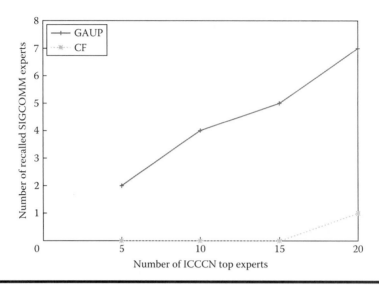

Figure 7.6 A comparison of GAUP and CF for finding experts in both *ICCCN* and *SIGCOMM*.

Table 7.11   Topic Drift of HITS Algorithm Using Three Different Seed Sets

20-Set	30-Set	31-Set
Donald F. Towsley	Donald F. Towsley	Wen Gao
James F. Kurose	James F. Kurose	Xilin Chen
Zhen Liu	Zhen Liu	Shiguang Shan
Weibo Gong	Krithi Ramamritham	Qingming Huang
Vishal Misra	Weibo Gong	Debin Zhao

■ *30-set.* We choose 10 new users together with 20-set, where some of new users are not in the field of networks.
■ *31-set.* We add a new user, Wen Gao, to 30-set.

Table 7.11 shows the top five experts found by the HITS algorithm for these three seed sets. The results of 20-set and 30-set only differ for one person, and all of them are well-known networking people. However, none of the results for 31-set is in the networking field. This is because Wen Gao and its neighboring nodes are well connected in the graph generated by HITS. As a result, the most highly ranked authorities and hubs deviate from the original topic of *SIGCOMM*.

This experiment demonstrates that the quality of HITS is highly dependent on the seed set. A direct comparison of HITS and our GAUP is difficult because GAUP does not need such seeds. We emphasize that GAUP is more reliable than HITS for finding experts because of the considerations for user preferences in the influence model.

Further Readings

Recommender Systems Handbook
http://www.springer.com/computer/ai/book/978-0-387-85819-7
F. Ricci, L. Rokach, B. Shapira, and P. B. Kantor, Eds., *Recommender Systems Handbook.*
New York, NY: Springer, 2011.
This book illustrates the technologies of recommendation systems behind well-known corporations such as Amazon, Yahoo!, Google, Microsoft, and AT&T.

Networks, Crowds, and Markets: Reasoning About a Highly Connected World
http://www.cs.cornell.edu/home/kleinber/networks-book/
D. Easley and J. Kleinberg. *Networks, Crowds, and Markets: Reasoning about a Highly Connected World, 1st ed.*, Cambridge University Press: New York, 2010.
This book covers a remarkable range of topics. By combining graph theory, probability and statistics, microeconomics, and facets of the social sciences, the book offers a broad new vision of how networks work.

WEKA
http://www.cs.waikato.ac.nz/ml/weka/
WEKA is an open source Java library that contains a collection of machine learning algorithms for data mining tasks.

SNAP
http://snap.stanford.edu/data/index.html
Stanford Large Network Dataset Collection (SNAP) provides different social network, road network, Web graph, and online community data. It is also the name for an open source library for processing large networks.

References

1. G. Linden, B. Smith, and J. York, Amazon.com recommendations: Item-to-item collaborative filtering, *IEEE Internet Computing*, vol. 7, no. 1, pp. 76–80, 2003.
2. A. Das, M. Datar, and A. Garg, Google news personalization: Scalable online collaborative filtering, in *Proceedings of the 16th International Conference on World Wide Web*, 2007, pp. 271–280.
3. T. Chen, W.-L. Han, H.-D. Wang, Y.-X. Zhou, B. Xu, and B.-Y. Zang, Content recommendation system based on private dynamic user profile, in *Proceedings of the Sixth International Conference on Machine Learning and Cybernetics*, 2007, pp. 2112–2118.
4. X. Su and T. M. Khoshgoftaar, A survey of collaborative filtering techniques, *Advances in Artificial Intelligence*, vol. 2009, 2009.

5. B. Sarwar, G. Karypis, J. Konstan, and J. Riedl, Item-based collaborative filtering recommendation algorithms, in *Proceedings of the 10th International Conference on World Wide Web*, 2001, pp. 285–295.

6. J. Wang, A. P. de Vries, and M. J. T. Reinders, Unifying user-based and item-based collaborative filtering approaches by similarity fusion, in *SIGIR '06*, 2006, pp. 501–508.

7. G.-R. Xue, C. Lin, Q. Yang, W. Xi, H.-J. Zeng, Y. Yu, and Z. Chen, Scalable collaborative filtering using cluster-based smoothing, in *SIGIR '05*, 2005, pp. 114–121.

8. D. Zhang, J. Cao, J. Zhou, M. Guo, and V. Raychoudhury, An efficient collaborative filtering approach using smoothing and fusing, in *ICPP*, 2009, pp. 558–565.

9. T. Zhang and V. S. Iyengar, Recommender systems using linear classifiers, *The Journal of Machine Learning Research*, vol. 2, pp. 313–334, 2002.

10. Y. Zhang and J. Koren, Efficient Bayesian hierarchical user modeling for recommendation system, in *SIGIR '07*, 2007, pp. 47–54.

11. Y. Koren, R. Bell, and C. Volinsky, Matrix factorization techniques for recommender systems, *IEEE Computer*, vol. 42, no. 8, pp. 42–49, 2009.

12. R. Salakhutdinov and A. Mnih, Probabilistic matrix factorization, in *NIPS*, 2008.

13. Comparison of feed aggregators. [Online]. Available from: http://en.wikipedia.org/wiki/Comparison_of_feed_aggregators last modified on April 1, 2016.

14. A. P. O'Riordan and M. O. O'Mahony, {InterSynd}: A web syndication intermediary that makes recommendations, in *iiWAS '08: Proceedings of the 10th International Conference on Information Integration and Web-based Applications & Services*, 2008.

15. J. J. Samper, P. A. Castillo, L. Araujo, C. J. J. Merelo, and F. Tricas, NectaRSS, an intelligent RSS feed reader, *Journal of Network and Computer Applications*, vol. 31, no. 4, pp. 793–806, 2008.

16. YUI Library, http://developer.yahoo.com/yui/. [Online]. Available from: http://developer.yahoo.com/yui/, accessed on March 7, 2016.

17. Google Reader Help. [Online]. Available from: http://www.google.com/support/reader/?hl=en, accessed on August 18, 2010.

18. PostRank, Inc., 2010. [Online]. Available from: http://www.postrank.com/, accessed on August 18, 2010.

19. C. D. Manning, P. Raghavan, and H. Schütze, *Introduction to Information Retrieval*, Cambridge University Press, Cambridge, UK, 2008.

20. H. Ma, I. King, and M. R. Lyu, Effective missing data prediction for collaborative filtering, in *Proceedings of the 30th Annual International ACM SIGIR Conference on Research and Development in Information Retrieval (SIGIR)*, 2007, pp. 39–46.

21. GroupLens Lab, "Social Computing Research at the University of Minnesota". http://grouplens.org, accessed on April 10, 2016.

22. J. L. Herlocker, J. A. Konstan, L. G. Terveen, and J. T. Riedl, Evaluating collaborative filtering recommender systems, *ACM Transactions on Information Systems (TOIS)*, vol. 22, no. 1, pp. 5–53, 2004.

23. R. Jin, J. Y. Chai, and L. Si, An automatic weighting scheme for collaborative filtering, in *SIGIR '04*, 2004, pp. 337–344.

24. T. Hofmann, Latent semantic models for collaborative filtering, *ACM Transactions on Information Systems (TOIS)*, vol. 22, no. 1, pp. 89–115, 2004.

25. D. M. Pennock, E. Horvitz, S. Lawrence, and C. L. Giles, Collaborative filtering by personality diagnosis: A hybrid memory and model-based approach, in *UAI '00*, 2000, pp. 473–480.

26. Y. Wang, G. Cong, G. Song, and K. Xie, Community-based greedy algorithm for mining top-K influential nodes in mobile social networks, in *KDD*, 2010, pp. 1039–1048.

27. M. Kimura and K. Saito, Tractable models for information diffusion in social networks, in *PKDD, LNAI 4213*, 2006, pp. 259–271.

28. J. Hartline, V. S. Mirrokni, and M. Sundararajan, Optimal marketing strategies over social networks, in *Proceedings of the 17th International Conference on World Wide Web (WWW)*. Beijing, China 2008, pp. 189–198.

29. H. Ma, H. Yang, M. R. Lyu, and I. King, Mining social networks using heat diffusion processes for marketing candidates selection, in *Proceeding of the 17th ACM Conference on Information and Knowledge Management (CIKM)*, 2008, pp. 233–242.

30. J. Leskovec, A. Krause, C. Guestrin, C. Faloutsos, J. VanBriesen, and N. Glance, Cost-effective outbreak detection in networks, in *KDD*, 2007.

31. C. S. Campbell, P. P. Maglio, A. Cozzi, and B. Dom, Expertise identification using email communications, in *CIKM*, 2003, pp. 528–531.

32. J. Zhang, M. S. Ackerman, and L. Adamic, Expertise networks in online communities: Structure and algorithms, in *Proceedings of the 16th International Conference on World Wide Web (WWW)*. Banff, Alberta, Canada, 2007, pp. 221–230.

33. D. Kempe, J. Kleinberg, and É. Tardos, Maximizing the spread of influence through a social network, in *Proceedings of the Ninth ACM SIGKDD International Conference on Knowledge Discovery and Data Mining (KDD)*, 2003, pp. 137–146.

34. P. Domingos and M. Richardson, Mining the network value of customers, in *Proceedings of the Seventh ACM SIGKDD International Conference on Knowledge Discovery and Data Mining (KDD)*, 2001, pp. 57–66.

35. M. Richardson and P. Domingos, Mining knowledge-sharing sites for viral marketing, in *Proceedings of the Eighth ACM SIGKDD International Conference on Knowledge Discovery and Data Mining (KDD)*, 2002, pp. 61–70.

36. W. Chen, Y. Wang, and S. Yang, Efficient influence maximization in social networks, in *Proceedings of the 15th ACM SIGKDD International Conference on Knowledge Discovery and Data Mining (KDD)*, 2009, pp. 199–208.

37. K. Saito, M. Kimura, K. Ohara, and H. Motoda, Learning continuous-time information diffusion model for social behavioral data analysis, in *Proceedings of the 1st Asian Conference on Machine Learning: Advances in Machine Learning (ACML)*, 2009, pp. 322–337.

38. K. Saito, M. Kimura, K. Ohara, and H. Motoda, Behavioral analyses of information diffusion models by observed data of social network, in *Proceedings of the 2010 International Conference on Social Computing, Behavioral Modeling, Advances in Social Computing Prediction (SBP10)*, 2010, pp. 149–158.

39. J. Tang, J. Sun, C. Wang, and Z. Yang, Social influence analysis in large-scale networks, in *Proceedings of the ACM SIGKDD International Conference on Knowledge Discovery and Data Mining*, 2009, pp. 807–816.

40. S. Deerwester, S. Dumais, G. Furnas, T. Landauer, and R. Harshman, Indexing by latent semantic analysis, *Journal of the American Society for Information Science*, vol. 41, no. 6, pp. 391–407, 1990.

41. J. Cadzow, SVD representation of unitarily invariant matrices, *IEEE Transactions on Acoustics, Speech and Signal Processing*, vol. 32, no. 3, pp. 512–516, 1984.

42. D. Gleich and L. Zhuko, SVD based term suggestion and ranking system, in *Proceedings of the Fourth IEEE International Conference on Data Mining*, pp. 391–394, 2004.

43. G. Cornuejols, M. Fisher, and G. Nemhauser, Location of bank accounts to optimize float, *Management Science*, vol. 23, no. 8, pp. 789–810, 1977.

44. G. Nemhauser, L. Wolsey, and M. Fisher, An analysis of the approximations for maximizing submodular set functions, *Mathematical Programming*, vol. 14, pp. 265–294, 1978.

45. J. M. Kleinberg, Authoritative sources in a hyperlinked environment, *Journal of the ACM*, vol. 46, no. 5, pp. 604–632, 1999.

46. K. Bharat and M. R. Henzinger, Improved algorithms for topic distillation in a hyperlinked environment, in *SIGIR*, 1998, pp. 104–111.

47. L. Li, Y. Shang, and W. Zhang, Improvement of HITS-based algorithms on web documents, in *Proceedings of the 11th International Conference on World Wide Web (WWW)*. Honolulu, Hawaii, USA, 2002, pp. 527–535.

Chapter 8

Case Studies

8.1 iCampus Prototype

The iCampus prototype consists of a number of pervasive applications designed to facilitate campus users in their daily activities. The prototype is based on our iShadow architecture [1]. In the following sections, we first describe the iShadow architecture and then discuss pervasive applications built on top of it.

8.1.1 iShadow: Pervasive Computing Environment

We developed a pervasive computing environment called Intelligent Shadow (iShadow) to provide minimal but flexible system support for adaptive pervasive applications [1]. The goals of iShadow include: (1) build a smart, pervasive computing platform, which offers fundamental functions to diverse applications and (2) provide customized user experience. Users personalize the services according to their preferences, which embodies the philosophy that pervasive computing is human-centric. By utilizing the basic services provided by iShadow, the system is capable of unobtrusively tracking the user and adjusting accordingly. The name iShadow illustrates the idea that our system accompanies a user like a shadow wherever the user goes, and it offers personalized services based on user context. Specifically, iShadow concentrates on the general architecture, offering flexible, low-level support to the pervasive applications through gracefully integrating human attention with computation. The key techniques of iShadow include:

- We design a lightweight user-shadow model to cater to context awareness and to track the user anytime, anywhere. Moreover, the user-shadow can be tailored to context and user preferences.

- We provide a scalable, distributed resource discovery mechanism that is based on a hash structure with two synchronization methods.
- We offer a potent context inference mechanism. iShadow employs the ontology to represent context and Bayesian networks for causal reasoning with which ontology does not cope well.

Figure 8.1 illustrates the hierarchical structure of iShadow, consisting of a devices layer, a middleware layer, and application layers. The devices layer provides a means for users to access services in pervasive computing environments. Computing devices include a PC, cell phone, and a personal digital assistant (PDA); sensor devices mainly contain devices that sense the physical environment; and network devices make sure the network is connected. These devices differ in terms of computation capability, memory, and human interface, leading to the heterogeneity in pervasive computing.

The middleware layer is crucial to iShadow and consists of resource management, user management, context awareness, message management, service management, and data management—where resource management and user management are two essential modules. We restrict the communication among middleware modules by means of a security control bus. A broad consensus of design and development has been reached from the existing architecture of the pervasive computing environment [2–5]. For instance, data management, service management, and message management are almost implemented in a parallel manner; whereas context representation using ontology is much more efficient [6,7]. We have designed the iShadow architecture similarly to the previous work. During the process, our architecture has gradually developed three distinctive characteristics: a lightweight user-shadow model in a user tracking module, scalable resource discovery in a resource discovery module, and a potent context inference mechanism in a context-awareness module.

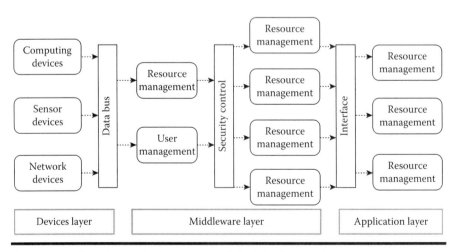

Figure 8.1 The hierarchical structure of the iShadow environment.

The application layer of iShadow refers to various applications, relying on the services provided by the middleware layer. In a sense, any application in specific smart spaces can be rapidly prototyped by calling the lower-level service. We cooperate with international collaborators in researching key pervasive computing issues by building a smart campus, smart car, and smart office using the iShadow architecture. Note that we focus on the salient characteristics of iShadow and omit the remainder modules of iShadow that have been extensively studied in previous systems (e.g., Gaia [8], EasyLiving [9], and Aura [10]).

8.1.2 Applications in iCampus Prototype

Based on the iShadow architecture, we have developed a number of pervasive applications for a campus environment. Figure 8.2 illustrates our iCampus prototype and three scenarios: student dorm area, office area, and work area for teaching. The user can access iCampus networks by PC, PDA, cell phone, or laptop across the campus. The prototype includes various services such as a map service, photo sharing service, search service, and daily work assisting service.

- *Map service.* This is deployed on our campus as an application of the campus guide. When a user visits the campus, the map service guides the user in navigating our campus. Map service varies with the user identification. If the user is a visitor to our campus, the map service will give detailed directions and provide some introductory information related to the user's location. For the users who often go to the university, the map service just shows some scenic pictures and the routes that the users are traversing.
- *Photo sharing service.* After a user takes some beautiful photos of a campus building, he is able to upload them to the photo sharing service, which facilitates the sharing of photos and the construction of a social network.
- *Search service.* The abundant resources of our university cannot be indexed by commercial search engines such as Google, Yahoo, and Baidu because of the access control and other security techniques. However, school resources are valuable to both students and faculty. We implemented a small-scale search engine, which only indexes the school resources. Our search engine can search according to the user preference, such as a profession and research interest.
- *Daily work assisting service.* This is an interesting service, designed to be used during a user's work day. When approaching his office or lab, the user wakes up his PC via a handheld device, such as a PDA or cell phone. Then, his PC will check his most used e-mail box; update his subscribed morning paper; synchronize the calendar among the PDA, the PC itself, and his Google calendar; and notify the user about new e-mails and appointments. When the time to leave work is approaching, the service notifies the user of the weather or about the bus according to the real-time situation.

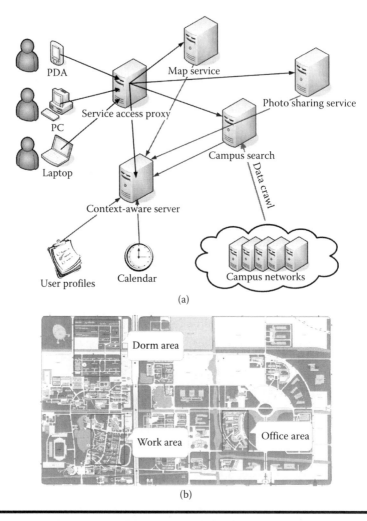

(a)

(b)

Figure 8.2 (a) iCampus architecture and (b) three scenarios of our campus, covering about 2,843,421 m².

We created an intelligent shadow for each user and utilized the shadow to track the user. When a user moves into or leaves a scenario, a resource discovery module will become aware of the change in λ period and notify the system, which makes some adaptations correspondingly. We extract the user preferences and current contexts from the map service, photo sharing service, calendar information, and the user's search behavior. Then, we use ontology or Bayesian networks to infer high-level contexts. Finally, the system will give the user recommendations or will execute actions. As a result, we hide the computing and greatly improve the user experience. Figure 8.3a through d is made up of snapshots of the map service, photo sharing service, search service, and the notification service, respectively.

(a)   (b)

(c)   (d)

Figure 8.3   The application screenshots of the iCampus prototype: (a) campus guide running at PDA; (b) campus guide running at Web; (c) school bus notification; and (d) morning paper notification.

8.2 IPSpace: An IPv6-Enabled Intelligent Space

A ubiquitous computing environment is a seamless fusion of the physical world and cyberspace, where people are able to access relevant information through any device, anytime, and anywhere in the social intelligent space. To realize this vision, a key problem to solve is how to make various types of devices, especially traditional nondigital equipment, able to connect to the pervasive space. We developed a prototype system called IPSpace, which is based on IPv6 protocol and enables traditional devices to communicate with cyberspace.

IPv6 is the next-generation network designed to replace IPv4. The 128-bit address space of IPv6 could supply an almost infinite address space and every device from mobiles to sensors could have an independent IP address. In additional, IPv6 has a stateless autoconfiguration protocol that is quite advantageous in deployment of large quantities of sensors without human intervention. However, IP connectivity is desired by WSNs but running a full-fledged IPv6 protocol is not suitable for sensors. In RFC 4944, IETF proposed 6LoWPAN [11], a first industry standard solution for IP communication in a low rate on an IEEE 802.15.4 link-layer. Specifically, 6LoWPAN is an adaptation layer between 802.15.4 and IPv6 and is used in our system for connecting various types of devices.

IPSpace faces several challenges. The first is resource management and discovery. Traditionally, applications built on sensor networks always work in a centralized architecture. Sensor devices send their data to target servers. Servers are responsible for processing these raw data and integrating them into different applications. External user requests could not reach end devices but could reach servers. With the 6LoWPAN, every device could be routed by an IP, and so a new distributed inter-operation mode is introduced in the personal networks. Changing from centralized operations to a distributed mode, how to find the service provided by the device, and how to manage these distributed services are important issues.

The second challenge is security. In the MAC layer 802.15.4 [12], the symmetric key AES algorithm is used for the encryption and is often implemented in the hardware. However, differing from previous threats, in the IP-enabled LR-WPANs, every device could be reached by an external IP address so there will be new attacks on the application layer; but 6LoWPAN does not specify any end-to-end security mechanism. Moreover, due to low-power and low computation capacity characteristics, a traditional security mechanism may not work directly, and software implementation is not acceptable for these tiny devices where complex hardware implementation contradicts low power consumption.

We propose an IP-enabled service LR-WPANs scheme with 6LoWPAN protocol. Specifically, we use Web services technology to make devices work as service providers. An XML format is used to enhance interoperability among devices. Finally, a multilayer authentication mechanism is used in our scheme to provide security.

8.2.1 IPSpace Architecture Overview

Figure 8.4 depicts an overview of the IPSpace architecture. There are three layers: the device layer, the service layer, and the security layer.

8.2.1.1 Device Layer

The device layer contains the end device, coordinator, and service grids. There are three types of devices defined in the 802.15.4 standard—the coordinator, the reduce function device (RFD), and the full function device (FFD). The coordinator is responsible for initiating a personal area network (PAN) and working as a gateway to other networks. There is exactly one coordinator in PAN. The RFD device is implemented in low-cost losing performance to router the frame. The FFD is

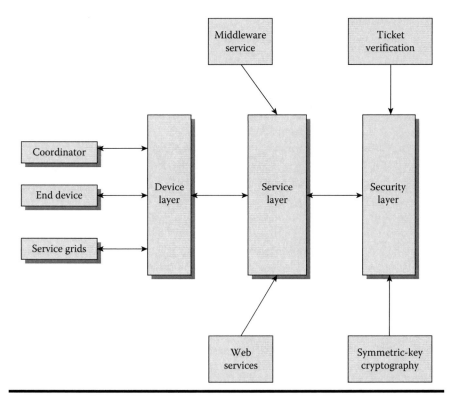

Figure 8.4   IPSpace architecture.

a device with complete stack and will often work as a router in the mesh topology. The end device is the basic element in IPSpace that has the sensor function and is responsible for collecting context information. Both the RFD and FFD devices can work as an end device. The coordinator is the initiator in PAN and has interfaces to connect to the Internet; it is a gateway router that sits between the IPv6 network and the 802.15.4-based LR-WPANs. Its calculation capability is more powerful than most end devices, and it has much more RAM space for storing routing information. In our architecture, the coordinator plays a role not only in maintaining a PAN but also in providing service registration functions for end devices to register their services with it. The coordinator, as a gateway router, will filter the external packet and ensure a reliable session between the external device and the internal device. In home use, a PC with an 802.15.4 transceiver can play the role of the coordinator as well. The coordinator is not capable of processing large quantities of service requests simultaneously. Clusters or high-performance mainframes compose service grids. Service grids establish repositories for services provided by sensors. Moreover, distributed services are integrated together to form large applications in service grids. Compared with the coordinator, service grids offer high reliability for large chunks of service requests and for user verification.

8.2.1.2 Service Layer

At the service layer, each device is a service provider in 6LoWPAN. However, different types of sensors are always in inconsistent data formats that result in poor exchangeability. XML language is a standardized and semantic data format that is platform-independent and self-described. Using XML in wireless sensor networks (WSNs) is a quite useful approach for supporting complex data management. Moreover, XML is a key element for Web services.

We use Web services technology for data exchange among devices in the service layer. With Web services, we convert a sensor function into a Web-service function, making it easy to be accessed by any Internet browsers. Considering the low-power characteristic of WSNs, running full-fledged Web services on devices is not practical. A lightweight middleware serves as an adaptation layer between Web services and the underlying communication framework. A function provided by sensors calls registration API from the middleware to notify the middleware that this function will work as a service. The middleware has a list of registered functions defined by our specified XML elements and sends these function descriptions in XML format to the coordinator when a device joins a 6LoWPAN. The coordinator will register these received function descriptions as services. The middleware of the coordinator supports translation from our specified service descriptions to Web services. The coordinator is a minimum service registration center for end devices in its domain. Service grids are centers of services from sensors. Using center management not only provides a more convenient approach for searching right services but also supports more management technologies for networks such as access control and load balancing.

8.2.1.3 Security Architecture

In security architecture, a multilayer authentication mechanism is proposed to address the network and application layer security that is not defined by IETF. The characteristics of 6LoWPAN and sensor devices require a light and scalable authentication mechanism. Although a traditional public-key cryptosystem works well in an IP network, it is not appropriate for WSNs. We adopt a multilayer, symmetric-key cryptography mechanism that is fast for those tiny devices and requires fewer resources than a public-key cryptosystem. We also use a ticket to build a secure connection between the user and services with which the cost for verification decreases.

Figure 8.5 shows the security architecture. The gateway servers between the Internet and PAN are called service grids. They are responsible for user authentication in the security architecture. They would check whether or not a user is valid. It is likely that different end devices offer different services. SSO [13] is applied to the architecture in order to reduce password fatigue from different user name and password combinations and to reduce time spent re-entering passwords for the same user.

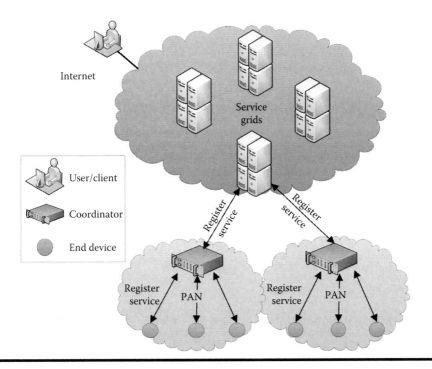

Figure 8.5 System security architecture.

User names and passwords should be stored in service grids' databases by administrators in advance. To avoid eavesdropping attacks, the client performs a one-way function (hash, usually) on the entered password, and this becomes the primary key (Kp) of the user.

The client sends a plain text message with the user ID and pre-authentication data to the service grids on behalf of the user. The pre-authentication data are a timestamp encrypted by Kp. It is used to verify that the user knows the password so that he/she could be trusted as a valid user. The service grids generates the Kp by hashing the password of the user found at the database and decrypts the timestamp to check if the client is a valid user.

If service grids are responsible for the user's state after a logon session, the service grids must keep in touch with the client. This could lead to the service grids becoming a bottleneck in the whole system. To solve this problem, a ticket called a TSSO is used to manage the user's state. In this case, TSSO is a long-term ticket that the client keeps. After a user obtains a TSSO, he/she could directly request services. The TSSO ticket is encrypted so that the client could not know its content.

After successfully logging on to the network, the user could request services in the PAN. The session key gained from the service grids is useful here. The user tells the coordinator which service he/she would like to use and shows the TSSO. Figure 8.6 shows the authentication process.

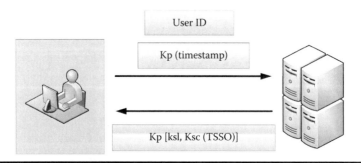

Figure 8.6 Process of user authentication.

8.2.2 Lightweight Embedded Web Services

Web services are XML-based protocols designed for addressing the interoperability among different applications and platforms and for delivering universal access to other applications. A sensor network itself is a data-centric network that is in accordance with service architecture.

8.2.2.1 Middleware Layer

Considering that a full set of Web services implemented on tiny sensor devices will cost too much CPU time, we built a lightweight, service middleware layer to implement service communication. The middleware defines several XML elements to describe methods working as services in a concise way comparing Web services and containing a CGI function to finish the HTTP transmission. Figure 8.7 depicts the service registration process. When an end node joins a 6LoWPAN, it first communicates with the coordinator, configures its IPv6 address, and obtains the coordinator address. After the configuration, the middleware sends the registration message of services provided by the sensor device to the coordinator in the XML format. The coordinator receives the service registration message, and the middleware layer of the coordinator will translate the service description to the Web services, creating the WSDL and SOAP definitions of the service. Finally, the coordinator will acknowledge to the sensor device denoting that the registration process is finished.

Figure 8.8 describes how our service module responds to the external request. In the service registration process, the coordinator has created the WSDL to describe the service, and the external user will get the WSDL to create a remote object and send the SOAP message to request service.

When a SOAP request comes in, the Web services module receives the request and redirects it to the service middleware. The coordinator provides the middleware with a full Web services stack for translating the Web services element to our middleware elements, even though it is not supported by end devices.

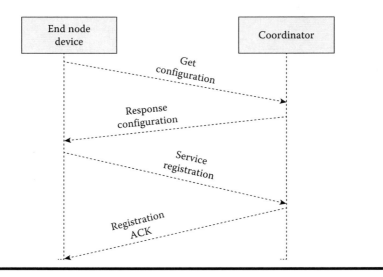

Figure 8.7   Service registration process.

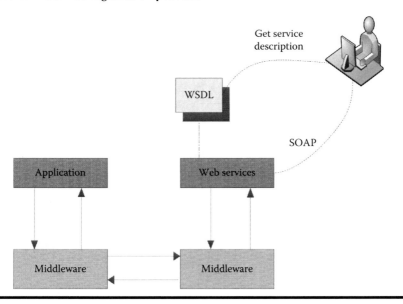

Figure 8.8   Service request process.

After the translation process finishes, the middleware sends the service request to the end devices via the coordinator. End devices receive the request and parse the service name and input parameters; then they provide the target service handler with the parameters. After obtaining the return value, the end node will send the result, wrapped by middleware, to the coordinator. The coordinator middleware decomposes the response message and sends the value to the user in SOAP.

8.2.2.2 Implementation

From a hardware perspective, current embedded sensor devices are suites that include a sensor, an MCU, and a radio transceiver. There are many bus interfaces among the chips, and sequence control is different among the chips. For the software viewer, embedded sensor devices are becoming more complex when writing programs implementing entire protocol suites, including 802.15.4, 6LoWPAN, IP stack, and so on. Fortunately, current CPU and memory power can also support a lightweight operating system. In our service solution, we choose Contiki [14] to run on the embedded devices. Contiki uses Protothreads [15], a lightweight stackless thread mechanism, to implement a multithreading function.

Compared with TinyOS [16], which is another reputable sensor operating system (OS), Protothreads does not need a stack to maintain the thread context state, which decreases the memory usage for context switching. Protothreads uses a short-type variable to store the line number it has executed, and this property makes the thread continue to run from its last point. Moreover, a short-type integer only takes 2 bytes of memory space to save context info. Multithreading is a good feature for sensor devices because it increases the throughput and CPU effective usage. Another advantage of Contiki is that it contains an adaptive communication stack for WSNs and supports 6LoWPAN and uIP [17] natively.

In our implementation, a service middleware supports the user-defined functions, working as the Web services for external requests. When programmers implement functions to fulfill requirements such as retrieving context temperature, they can focus on implementing the application logic. We provide an underlying communication framework in the service middleware, and programmers just need to write some configurations and invoke some APIs from the middleware. How to publish the function and response external requests are left to the service middleware. An example is depicted in Figure 8.9. Function *getTemp* is written to obtain the surrounding temperature. There is no need for the programmer to write a network packet transmission code in it because this will be done by our middleware.

To expose the *getTemp* function as a service, we first write a services configuration XML file to define functions that will work as the service. Programmers call the API REGISTER_SERVICE to notify the middleware about the existence of the function. After the function registration, the middleware will create a handler for this function. The middleware will create a table to store the key value of the function name and handler. When an external SOAP request arrives at the coordinator, the middleware on the coordinator translates the SOAP request to our defined service format and sends this to the end node device. When the middleware receives the request, it will extract the service name from the request and look up the table to find the right handler to respond to the request. The handler is responsible for invoking the responding function and responding to the request with the return value.

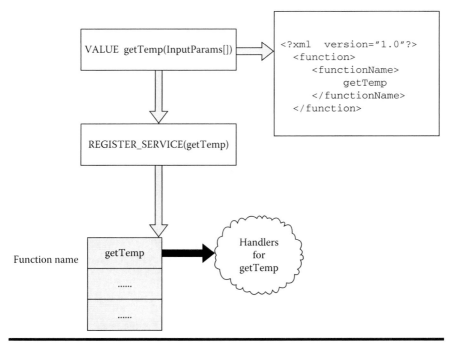

Figure 8.9   Example of middleware.

8.2.3 *Experiment*

We evaluate our IPSpace service module on the testbed of Raven kits [18] produced by Atmel. Raven is an evaluation and starter kit for demonstration of 6LoWPAN applications. The Raven kits comprise two AVR Raven boards and a USB stick. All devices have a 2.4 GHz transceiver for sending and receiving a wireless signal. The two AVR Raven boards are used as RFD devices, and the USB stick connects to a computer by the USB interface. Connecting with the USB stick, the computer works as the coordinator in this 6LoWPAN network. Raven kits are constructed in a default star topology, and all data packets need to pass through the coordinator. Memory is a key constraint for tiny devices. Take the Raven board as an example Atmel 1284p, the main processor unit on the board, has a 16 KB SRAM chip memory. Implementing service onboard requires a TCP stack, and we use the uIP for decreasing memory usage.

Figure 8.10 shows the example messages used in our experiment. In Table 8.1, we show the memory usage increase in each part. The base example has a primitive *getTemp* function that reads the temperature. In the base example, there is no networking code so the temperature could not be sent to other devices. We use the lightweight middleware to make the *getTemp* function work as a service. The code is stored in the text segment and the text segment is stored in ROM (128 KB on

```
<?xml version="1.0"?>
  <function>
    <functionName>
      getTemp
    </functionName>
  </function>
```

```
<?xml version="1.0"?>
  <Response>
    <functionName>
      getTemp
    </functionName>
    <ReturnValue>
      27
    </ReturnValue>
  </Response>
```

Service definition   Response message

Figure 8.10 Example messages.

Table 8.1 Code Size and Memory Usage

	Text (byte)	Data (byte)	Bss (byte)
Base example	39,866	1111	9370
After service translation	44,958	1407	9955

Table 8.2 Query Latency

	Data Length (byte)	Latency (second)
Request message	108	3.7
Registration message	79	3.1

Raven boards). The data segment and Bss segment are stored in SRAM. Compared with the base example, after translating into service, there are 881 bytes more memory usage including the data and Bss segments. The whole memory usage is 11,362 bytes less than 16 KB.

Table 8.2 shows the latency of the request message and the registration message from our example. The data length item is the length of the example XML message in the application layer. The latency is comprised of the MCU process time and the wireless transmission time. The latency is acceptable for most of the applications.

Further Readings

Urban Computing at Microsoft Research Asia
http://research.microsoft.com/en-us/projects/urbancomputing/
Urban computing acquires and analyzes large and heterogeneous data generated by a diversity of sources in urban spaces to tackle the major issues that cities face—such as air pollution, increased energy consumption, and traffic congestion.

Interaction Design Foundation: Wearable Computing
https://www.interaction-design.org/encyclopedia/wearable_computing.html
This section covers wearable computing, which is the study or practice of inventing, designing, building, or using miniature body-borne computational and sensory devices.

Gabriel: Wearable Cognitive Assistance Using Cloudlets
http://elijah.cs.cmu.edu/
This project at Carnegie Mellon University introduces cloudlet as a new architectural element that arises from the convergence of mobile computing and cloud computing.
This enabling technology supports new cognitive assistance applications that will
seamlessly enhance a user's ability to interact with the real world.

References

1. D. Zhang, H. Guan, J. Zhou, F. Tang, and M. Guo. iShadow: Yet another pervasive computing environment. In *Proceedings of International Symposium on Parallel and Distributed Processing with Applications (ISPA)*, pp. 261–268, 2008.
2. R. Masuoka, Y. Labrou, B. Parsia, and E. Sirin. Ontology-enabled pervasive comput-ing applications. *IEEE Intelligent Systems*, 18(5): 68–72, 2003.
3. D. Garlan, D. Siewiorek, A. Smailagic, and P. Steenkiste. Project aura: Toward dis-traction-free pervasive computing. In *IEEE Pervasive Computing*, pp. 22–31, 2002.
4. K. Henricksen, J. Indulska, and A. Rakotonirainy. Modeling context information in pervasive computing systems. In *Proceedings of the 1st International Conference on Pervasive Computing*, pp. 167–180, 2002.
5. G. Banavar and A. Bernstein. Software infrastructure and design challenges for ubiqui-tous computing applications. *Communications of the ACM*, 45(12): 92–96, 2002.
6. H. Chen, T. Finin, and A. Joshi. An ontology for context-aware pervasive computing environments. *The Knowledge Engineering Review*, 18(3): 197–207, 2004.
7. X. Wang, D. Zhang, T. Gu, and H. Pung. Ontology based context modeling and reasoning using OWL. *PerCom Workshops*, pp. 18–22, 2004.
8. M. Romań, C. K. Hess, R. Cerqueira, and A. Ranganathan. Gaia: A middleware infra-structure to enable active spaces. In *IEEE Pervasive Computing*, pp. 74–83, 2002.
9. B. Brumitt, B. Meyers, J. Krumm, A. Kern, and S. A. Shafer. EasyLiving: Technologies for intelligent environments. In *Proceedings of 2nd International Symposium on Handheld and Ubiquitous Computing, HUC 2000*, pp. 12–29, 2000.
10. J. P. Sousa and D. Garlan. Aura: An architectural framework for user mobility in ubiquitous computing environments. In *Proceedings of 3rd IEEE/IFIP Conference on Software Architecture*, pp. 29–43, 2002.
11. Transmission of IPv6 Packets over IEEE 802.15.4 Networks. 2007. Available from: http://www.ietf.org/rfc/rfc4944.txt, accessed on May 3, 2015.
12. IEEE standard 802.15.4-2006. Wireless Medium Access Control (MAC) and Physical Layer (PHY) Specifications for Low-Rate Wireless Personal Area Networks (WPANs). 2006.
13. Dae-Hee Seo, Im-Yeong Lee, Soo-Young Chae and Choon-Soo Kim, *Single sign-on authentication model using MAS (multiagent system)*, In *Proceedings of IEEE Pacific Rim Conference on Communications, Computers and signal Processing (PACRIM)*, August 28–30, pp. 692–695, 2003.

14. A. Dunkels, B. Grönvall, and T. Voigt. Contiki—A lightweight and flexible operating system for tiny networked sensors. In *IEEE EmNets 2004*.

15. A. Dunkels, O. Schmidt, T. Voigt, and M. Ali. Protothreads: Simplifying event-driven programming of memory-constrained embedded systems. In *ACM SenSys 2006*.

16. TinyOS. Available from: http://www.tinyos.net/, accessed on December 20, 2009.

17. uIP. Available from: http://www.sics.se/~adam/uip/index.php/Main_Page, accessed on December 20, 2009.

18. ATAVRRZRAVEN 2.4 GHz Evaluation and Starter Kit. Available from: http://www.atmel.com/dyn/Products/tools_card.asp?tool_id=4291, accessed on December 20, 2009.

Index